Fire Engineering's Handbook
for Firefighter I & II
STUDY GUIDE
2019 Update

FIRE ENGINEERING'S HANDBOOK FOR FIREFIGHTER I & II STUDY GUIDE: 2019 UPDATE

Copyright © 2019 by
Fire Engineering Books & Videos
1421 South Sheridan Road
Tulsa, Oklahoma 74112-6600 USA

Phone: 918.835.3161
Fax: 918.831.9555
800.752.9764
+1.918.831.9421
info@fireengineeringbooks.com
www.fireengineeringbooks.com

Senior Vice President Eric Schlett
Operations Manager Holly Fournier
Sales Manager Josh Neal
Managing Editor Mark Haugh
Production Manager Tony Quinn
Technical Editor Glenn Corbett
Development Editor Chris Barton
Book Designer Sheila Brock
Cover Designer Beth Rose

All rights reserved. No part of this book may be reproduced, stored in a retrieval system, or transcribed in any form or by any means, electronic or mechanical, including photocopying and recording, without the prior written permission of the publisher.

Printed in the United States of America

1 2 3 4 5   23 22 21 20 19

The recommendations, advice, descriptions, and methods in this book are presented solely for educational purposes. The authors and publisher assume no liability for any loss or damage resulting from the use of any material in this book. Users of this book should follow all standard operating procedures and safety procedures for the user's department or training facility, and use all appropriate protective clothing, gear, and equipment relevant to the activities described herein. The reader assumes all risks associated with following the instructions contained in this book, and should take every precaution to avoid potential hazards.

CONTENTS

FIREFIGHTER I

1
THE TRADITIONS AND MISSION OF THE FIRE SERVICE ... 1
Questions ... 1
Answer Key ... 3

2
FIRE SERVICE HISTORY ... 2
Questions ... 5
Answer Key ... 14

3
FIRE DEPARTMENT ORGANIZATION
Questions ... 17
Answer Key ... 22

4
FIRE DEPARTMENT COMMUNICATIONS
Questions ... 25
Answer Key ... 34

5
FIRE BEHAVIOR
Questions ... 37
Answer Key ... 43

6
FIRE EXTINGUISHERS
Questions ... 45
Answer Key ... 58

7
BUILDING CONSTRUCTION
Questions ... 61
Answer Key ... 69

8
ROPES AND KNOTS
Questions ... 71
Answer Key ... 80

9
PERSONAL PROTECTIVE EQUIPMENT
Questions ... 83
Answer Key ... 89

10
SELF-CONTAINED BREATHING APPARATUS
Questions ... 91
Answer Key ... 99

11
FIREFIGHTING TOOLS
Questions 101
Answer Key 110

12
FORCIBLE ENTRY
Questions 113
Answer Key 122

13
LADDERS
Questions 125
Answer Key 136

14
VENTILATION
Questions 139
Answer Key 144

15
WATER SUPPLY AND HOSE
Questions 147
Answer Key 160

16
FIRE STREAMS
Questions 163
Answer Key 172

17
FIREFIGHTER SAFETY AND SURVIVAL
Questions 175
Answer Key 184

18
VEHICLE FIRES
Questions 187
Answer Key 196

19
SEARCH AND RESCUE
Questions 198
Answer Key 207

20
BASIC FIRE ATTACK
Questions 209
Answer Key 223

21
SALVAGE AND OVERHAUL
Questions 227
Answer Key 233

22
EMERGENCY MEDICAL RESPONSE
Questions 235
Answer Key 240

FIREFIGHTER II

23
THE INCIDENT COMMAND SYSTEM
- Questions 243
- Answer Key 249

24
ADVANCED COMMUNICATIONS
- Questions 251
- Answer Key 254

25
PRE-INCIDENT PLANNING
- Questions 256
- Answer Key 264

26
FIRE PROTECTION SYSTEMS
- Questions 267
- Answer Key 276

27
ADVANCED FIRE ATTACK
- Questions 278
- Answer Key 299

28
ORIGIN AND CAUSE INVESTIGATION
- Questions 291
- Answer Key 295

29
FIRE PREVENTION AND PUBLIC EDUCATION
- Questions 297
- Answer Key 301

30
VEHICLE EXTRICATION
- Questions 303
- Answer Key 309

31
SUPPORT OF TECHNICAL RESCUE TEAMS
- Questions 311
- Answer Key 317

The Traditions and Mission of the Fire Service

by Rick Lasky

1. _____ is called the Father of the American Fire Service
   a. George Washington
   b. Thomas Jefferson
   c. Benjamin Franklin
   d. John Adams

2. In the early days of the fire service, becoming a firefighter was important to those who wanted to:
   a. make more money
   b. become a family
   c. climb politically or socially
   d. have an exciting career

3. Safety is the job of:
   a. the safety officer
   b. every firefighter
   c. the fire chief
   d. the police

4. Wetting down new apparatus is an example of a fire service:
   a. maintenance chore
   b. superstition
   c. training technique
   d. tradition

5. The quote "I have no ambition in the world but one, and that is to be a fireman" is attributed to:
   a. Hugh Halligan
   b. Edward Croker
   c. William Tweed
   d. Benjamin Franklin

6. In the early days of firefighting, these were used to determine who had the best fire company:
   a. musters and contests
   b. speeches and voting
   c. fundraising events

7. The job of shouting to clear the streets in front of responding fire apparatus was given to a runner, also known as a:
   a. Yeller
   b. Rookie
   c. Johnnie
   d. Torchbearer

8. The new firefighter should understand the need and importance of:
   a. company logos
   b. insignias
   c. tools
   d. all of the above

9. Which one of the following can legitimately report that a fire has been extinguished?
   a. police officers
   b. firefighters
   c. elected officials
   d. civilians

10. Your uniform represents the fire service's:
    a. historical importance
    b. organizational culture
    c. heritage

CHAPTER 1 ANSWER KEY

Question #	Answer	Page #
1	c	2
2	c	5
3	b	5
4	d	5
5	b	3
6	a	2
7	d	2
8	d	5
9	b	7
10	c	5

Fire Service History

by Don Cannon, updated by Glenn Corbett

1. In Rome, the Emperor Augustus is credited with creating the first public force of firefighters known as:

    a. Pompier
    b. Vigiles
    c. Knights
    d. Protectors

2. When the ship *Tyger* burned in 1613, it forced the crew to establish a settlement in what is now:

    a. Manhattan
    b. Boston
    c. Montreal
    d. Newport

3. In 1648, Peter Stuyvesant, governor of New Amsterdam, created a night patrol as a fire prevention measure, known as the:

    a. Protectors
    b. Patrolers
    c. Rattle Watch
    d. Night Watch

4. This fire in 1666 destroyed 80% of which city and influenced fire prevention measures in the building of Philadelphia?

    a. Great Fire of Rome
    b. Great Fire of London
    c. Great Fire of Paris
    d. Great Fire of Athens

5. America's first volunteer fire brigade, organized in Philadelphia in 1736 was called the:

    a. Alpha Fire Company
    b. Pioneer Fire Company
    c. Franklin Fire Company
    d. Union Fire Company

6. The Friendly Society of Charleston and the Philadelphia Contributership were two of the first _____ found in America.
   a. ladies auxiliaries
   b. volunteer fire companies
   c. career fire companies
   d. fire insurance companies

7. _____ were wood or iron plaques placed on buildings to identify the insurer of the property.
   a. Fire marks
   b. Property markers
   c. Protection marks
   d. Assurance marks

8. In the _____, Thomas Jefferson accused the English King of using fire as a weapon of terror.
   a. *Articles of Confederation*
   b. *Federalist Papers*
   c. *Declaration of Independence*
   d. *Philadelphia Gazette*

9. After a series of major fires in the years following the American Revolution, this city undertook a major water supply project that became a model for other cities:
   a. New York
   b. Philadelphia
   c. Boston
   d. Charleston

10. Early American hand-pumped fire engines with solid-metal play pipes were called:
    a. gooseneck style
    b. swan-neck style
    c. bent-pipe style
    d. squirrel-tail style

11. In the early 1800s, some engine companies used copper-riveted hose made from:
    a. cloth and resin
    b. metal
    c. cotton
    d. leather

12. On April 19, 1863, much of this western city was destroyed, causing the city to enact an ordinance requiring all new buildings to be made of brick or stone.
    a. San Francisco
    b. Seattle
    c. Denver
    d. Los Angeles

13. Firefighters reaching "exempt" status meant that they were forever excused from:
    a. fire duty
    b. jury and militia duty
    c. paying property taxes
    d. firehouse cleaning duties

14. This New York–based politician began his controversial career as a foreman of Americus Engine 6 in New York City.
    a. Fiorello LaGuardia
    b. Alfred E. Baker
    c. William M. Tweed
    d. Thomas Dewey

15. James Braidwood is credited with organizing the first effective paid fire department during 1883 in what city?
    a. London
    b. Cincinnati
    c. Boston
    d. Dublin

16. In 1852, Boston became the first city to install a:
    a. water cistern system
    b. fire hydrant system
    c. telephone alarm system
    d. street box fire alarm system

17. In 1853, this city became the first to employ a fully paid fire department:
    a. Pittsburgh
    b. Cincinnati
    c. Columbus
    d. New Orleans

18. During the Civil War, members of army regiments made up of firefighters were known as
    a. The Fire Soldiers
    b. The Fire Corps
    c. The Fire Zouaves
    d. The Fire Brigades

19. In 1863 members of New York City's Black Joke Engine 33 were said to have participated in the infamous
    a. Battle at Gettysburg
    b. Battle at Antietam
    c. Five Points Gang Wars
    d. Draft Riots

20. In 1865 the Metropolitan Fire Department took the place of volunteers in which city?
    a. Memphis
    b. New York
    c. Washington, DC
    d. San Francisco

21. In 1870, the New York City Fire Department was created under which legislation?
    a. NYC Terms of Organization
    b. Articles of Founding
    c. City Charter
    d. Tweed Charter

22. In 1873, Boston put into service the first American
    a. steam-powered fireboat
    b. steam-powered fire pumper
    c. fire department ambulance
    d. fire chief's car

23. Daniel Hayes of the San Francisco Fire Department is credited with developing the first successful
    a. steam-driven pump
    b. vehicle-mounted warning siren
    c. aerial extension ladder truck
    d. cellar pipe

24. Fire Prevention Week is held in early October of each year to commemorate which of the following?
    a. Baltimore Fire of 1904
    b. Great Chicago Fire
    c. San Francisco Fire and Earthquake
    d. Great Philadelphia Fire

25. In October 1871, this prairie fire burned 1.5 million acres in Wisconsin and Michigan and killed up to 2,500 people.
    a. Peshtigo
    b. Billings
    c. Green Bay
    d. Mackinaw

26. This organization was established in 1896 to standardize codes for sprinklers and electrical systems:
    a. National Fire Prevention Association
    b. National Fire Protection Association
    c. Insurance Services Office
    d. Underwriters Laboratories Inc.

27. The Baltimore Fire of 1904 was the impetus for the creation of a new
    a. sprinkler code
    b. electrical code
    c. fire prevention code
    d. model building code

28. The 1911 Triangle Shirtwaist fire led to improved laws for
    a. apartment buildings
    b. transportation facilities
    c. high-occupancy structures
    d. medical facilities

29. This 1906 event is considered to be the largest fire in American history:
    a. Great Chicago Fire
    b. Portland Fire
    c. San Francisco Earthquake and Fire
    d. Great Maine Forest Fire

30. By the year _____, motorized apparatus had replaced horse-drawn apparatus.
    a. 1919
    b. 1920
    c. 1921
    d. 1922

31. In July 1916, this fire and detonation of two million pounds of explosives in New Jersey occurred on
    a. Liberty Island
    b. Black Tom Island
    c. Long Beach Island
    d. Adams Island

32. In 1919 race riots in this city burned down almost 1,300 buildings.
    a. Los Angeles, California
    b. Tulsa, Oklahoma
    c. Phoenix, Arizona
    d. Seattle, Washington

33. A fire and explosion at the New London, Texas, Junior-Senior High School led to this safety requirement:
    a. the addition of mercaptan to odorize natural gas
    b. fire evacuation drills
    c. installation of sprinkler systems
    d. installation of fire alarm systems

34. In November 1942, this Boston fire killed 492 people:
    a. fire at Fenway Park
    b. Iroquois Theater Fire
    c. Cocoanut Grove Night Club
    d. Boston Dockyard Fire

35. In 1944 this Hartford, Connecticut, fire killed 167 people:
    a. Vendome Hotel Fire
    b. Iroquois Theater Fire
    c. Ringling Brothers & Barnum and Bailey Circus Tent Fire
    d. Hartford School Fire

36. In 1947 the ship *Grandcamp,* loaded with ammonium nitrate fertilizer, caught fire and blew up, killing 26 firefighters in which city?
    a. Houston, Texas
    b. Texas City, Texas
    c. Vicksburg, Mississippi
    d. San Diego, California

37. Commonly referred to as the "War Years," by urban firefighters, this period of increased fire activity took place during
    a. 1930s–1940s
    b. 1980s–1990s
    c. 1950s–1960s
    d. 1960s–1970s

38. This 1973 report by the National Commission on Fire Prevention and Control noted that the American fire service had expanded rapidly beyond fire prevention, suppression, and investigation.
    a. Burning America
    b. The National Fire Report
    c. The American Fire Report
    d. America Burning

39. In 1973 Arlington County, Virginia, made history by hiring the
    a. first female fire chief
    b. first Hispanic American fire chief
    c. first career woman firefighter
    d. first Native American firefighter

40. The first African American firefighter appointed to the Los Angeles County Fire Department occurred in
    a. 1953
    b. 1956
    c. 1959
    d. 1962

41. This type of fire is a growing hazard in many regions of the United States:
    a. wildland fires
    b. railway shipping fires
    c. industrial chemical fires
    d. computer wiring fires

42. This 1994 Colorado fire killed 14 wildland firefighters:
    a. Mann Gulch
    b. South Canyon Glenwood Springs
    c. Pueblo Fire
    d. Mendocino Forest Fire

43. The Incident Management System evolved out of an incident command model known as
    a. NFIRS
    b. ISCOPE
    c. FIRESCOPE
    d. LUNAR

44. In 1977 the Jacksonville, Florida, Fire Department established the first formal
    a. fire department hazardous material response team
    b. SCBA team
    c. paramedic engine companies
    d. rescue companies

45. In 1989 FEMA established the _____ to coordinate emergency services during major disasters.
    a. State Urban Search and Rescue System
    b. Citizens Emergency Response Teams
    c. Disaster Mitigation Response System
    d. National Urban Search and Rescue System

46. The 1910 Los Angeles Times building fire, the 1922 Wall Street bombing, and the 1964 Mississippi church fires are all considered to be
    a. not historically significant
    b. terrorist acts
    c. tied to labor unrest
    d. caused by foreign actors

47. After the attacks on the World Trade Center in September 2001, the role of the retired fire boat *John Jay Harvey* was significant because:
    a. It supplied the only firefighting water to the scene for 96 hours.
    b. It evacuated many people to New Jersey.
    c. It evacuated many people to Brooklyn.
    d. It was used as triage station for injured persons.

48. The greatest loss of firefighters to ever occur, the World Trade Center attack killed _____ FDNY firefighters.

    a. 143
    b. 243
    c. 343
    d. 443

49. As a thank you to the New York City Fire Department for a donation made in 1867, this South Carolina city raised funding to purchase a tiller ladder truck, to be used by the FDNY after the September 11, 2001, terrorist attack.

    a. Myrtle Beach
    b. Columbia
    c. Raleigh
    d. Anderson

50. A pumper dubbed the Spirit of Louisiana, donated to the FDNY, would later serve as a replacement for a damaged apparatus in which fire department?

    a. Nashville
    b. Mobile
    c. Biloxi
    d. New Orleans

CHAPTER 2 ANSWER KEY

Question #	Answer	Page #	Question #	Answer	Page #
1	b	11	31	b	21
2	a	12	32	b	21
3	c	12	33	a	21
4	b	12	34	c	22
5	d	12	35	c	22
6	d	12	36	b	22
7	a	12	37	d	22
8	c	14	38	d	23
9	b	14	39	c	23
10	a	14	40	a	23
11	d	14	41	a	24
12	c	15	42	b	24
13	b	15	43	c	24
14	c	16	44	a	24
15	a	16	45	d	24
16	d	17	46	b	24
17	b	17	47	a	25
18	c	17	48	c	25
19	d	17	49	b	26
20	b	18	50	d	26
21	d	18			
22	a	19			
23	c	19			
24	b	19			
25	a	19			
26	b	19			
27	d	19			
28	c	20			
29	c	20			
30	d	20			

Fire Department Organization

by John Best

3

1. Which major organization closely associated with the fire department provides and advocates on consensus codes, standards, research, training, and education?

    a. National Fire Protection Association (NFPA)

    b. International Association of Fire Chiefs (IAFC)

    c. International Association of Fire Fighters (IAFF)

    d. National Volunteer Fire Council (NVFC)

2. Fire companies are generally organized by:

    a. equipment carried

    b. geographic orientation

    c. function of the apparatus

    d. staffing requirements

3. This is a part of a department's master plan that also includes the department's vision, values, goals, and objectives:

    a. rules and regulations

    b. mission statement

    c. standard operating procedures

    d. table of organization

4. Regardless of whether the fire department is career or volunteer, a large complex entity, or a small municipal single-station organization, this is the primary position and resource in the fulfillment of the associated missions:

    a. the firefighter

    b. the apparatus

    c. the administration

    d. the facilities

5. This is a company assigned to deliver water, deploy hoselines, and execute other associated tactics to extinguish fires:
    a. engine company
    b. ladder company
    c. rescue company
    d. squad company

6. This company is assigned to accomplish search and rescue, perform ventilation and forcible-entry operations, secure utilities, and perform salvage or overhaul at fire and other emergency scenes:
    a. engine company
    b. ladder company
    c. rescue company
    d. combination company

7. A vehicle designed primarily for transporting (pickup, transporting, and delivering) water to fire emergency scenes to be applied by other vehicles or pumping equipment is known as a _____.
    a. quint
    b. squad
    c. tender
    d. wildland unit

8. Generally, the distinction between Firefighter I and Firefighter II is:
    a. amount of time in the service
    b. amount of in-service training a firefighter has had
    c. level of supervision required and ability to operate independently
    d. type of department the firefighter serves (career, volunteer, etc.)

9. Fire departments comprise multiple levels of supervision. The level of supervision depends on all of the following except:
    a. experience level of the subordinates
    b. size of the department
    c. authority governing the department or given to the supervisor
    d. responsibility of the organization.

10. I am responsible for the supervision of the personnel and resources associated with multiple units/stations or organizational disciplines within the fire department. Who am I?
    a. company officer
    b. battalion or district chief
    c. deputy or assistant chief
    d. fire commissioner or fire chief

11. One of my duties is to maintain, develop, and provide leadership for a risk management program to reduce department injuries and property damage. Who am I?
    a. company officer
    b. battalion or district chief
    c. deputy or assistant chief
    d. fire commissioner or fire chief

12. I am a station-level supervisor and must be familiar with departmental operations, as well as safety and administrative procedures. Who am I?
    a. company officer
    b. battalion or district chief
    c. deputy or assistant chief
    d. fire commissioner or fire chief

13. I am responsible for the supervision of the personnel and resources associated with a major functional discipline (training, emergency medical services, community risk reduction, etc.) within the fire department. Who am I?
    a. company officer
    b. battalion or district chief
    c. deputy or assistant chief
    d. fire commissioner or fire chief

14. This level of emergency medical service responder is considered to be the minimum level of certification required to function on an ambulance.
    a. first responder
    b. basic emergency medical technician (EMT-B)
    c. intermediate emergency medical technician (EMT-I)
    d. paramedic (PM)

15. Emergency medical response personnel operate under the auspices of:
    a. deputy chief
    b. hospital liaison
    c. medical director
    d. communications director

16. Advanced life support (ALS) systems are provided by:
    a. first responder
    b. basic emergency medical technician (EMT-B)
    c. intermediate emergency medical technician (EMT-I)
    d. paramedic (PM)

17. Most public safety answering points (PSAP) are facilitated by this agency where calls for assistance are initially received and then transferred to the fire department for dispatch:
    a. department communication liaisons
    b. municipal emergency coordinators
    c. local emergency medical services
    d. local law enforcement

18. These are typically used by the fire department to document the "how-to" in a consistent manner, so that all involved understand what is expected of them and their co-workers and how to accomplish it:
    a. regulations
    b. policies and procedures
    c. bylaws
    d. mutual aid plans

19. These documents generally have enforcement or compliance requirements:
    a. regulations
    b. policies and procedures
    c. bylaws
    d. mutual aid plans

20. _____ typically address operational matters such as hose loads, water supply, high-rise operations, and safety requirements:
    a. policies
    b. standard operating procedures
    c. regulations
    d. bylaws

21. _____ are more often associated with administrative matters such as time and attendance and personnel matters and are generally promulgated from management or senior staff.
    a. policies
    b. standard operating procedures
    c. regulations
    d. bylaws

22. These agreements between fire departments involve reciprocal assistance under a prearranged plan:
    a. reciprocal aid
    b. automatic aid
    c. mutual aid
    d. standardized move-up

23. This type agreement is a plan developed by two or more fire departments for the immediate joint response to incidents.
    a. reciprocal aid
    b. automatic aid
    c. mutual aid
    d. standardized move-up

24. The Federal Emergency Management Agency (FEMA), the United States Fire Administration (USFA), and the National Fire Academy are all resources provided by:
    a. International Association of Fire Chiefs
    b. Department of Homeland Security
    c. International Association of Firefighters
    d. National Fire Protection Association

25. Whether enforced at the state or federal level, these workplace requirements apply to volunteer and career fire departments.
    a. OSHA
    b. NIOSH
    c. ISO
    d. NFPA

CHAPTER 3 ANSWER KEY

Question #	Answer	Page #
1	a	29
2	c	30
3	b	30
4	a	31
5	a	32
6	b	33
7	c	34
8	c	34
9	a	36
10	b	37
11	d	38
12	a	36
13	c	37
14	b	38
15	c	39
16	d	39
17	d	40
18	b	42
19	a	42
20	b	42
21	a	42
22	c	43
23	b	44
24	b	45
25	a	45

Fire Department Communications

by Charles Jennings with Alan Young

1. Communications centers of all sizes must conform to standards required by
    a. Federal Emergency Management Association (FEMA)
    b. National Fire Protection Association (NFPA)
    c. National Institute of Occupational Safety and Health (NIOSH)
    d. Insurance Services Organization (ISO)

2. A public safety communications center must be built to conform with what NFPA standard pertaining to how buildings are designed and how notifications are to be sent?
    a. NPFA 901
    b. NFPA 1221
    c. NFPA 911
    d. NFPA 1300

3. A person trained and responsible for receiving and dispatching all nonemergency and emergency calls for service is known as a(n) _____.
    a. telecommunicator
    b. facilitator
    c. communications liaison
    d. emergency operator

4. In older communication systems, card systems were used to:
    a. Keep track of the companies in the field.
    b. Document personnel on duty.
    c. Determine a predesignated response to an emergency.
    d. Maintain the status of resources available to an incident.

5. The purpose of this type dispatching system is to have the ability to enter a call location, identify the type of incident the units will be responding to, and recommend the appropriate equipment for response to the identified incident type.

    a. automatic dispatching capability (ADC)

    b. complete automated dispatching (CAD)

    c. computer-enabled communications (CAC)

    d. computer-aided dispatch (CAD)

6. In the United States, the universal emergency number for all telephone services is:

    a. 4-1-1

    b. E-9-1-1

    c. 9-1-1

    d. 0

7. This type system provides enhanced equipment and database information that allows the telecommunicator to see the phone number and address of the caller on a display screen at the communications center.

    a. enhanced 9-1-1

    b. basic 9-1-1

    c. dual phase 9-1-1

    d. displayed 9-1-1

8. The largest drawback with wireless 9-1-1 calling is:

    a. The call may not go to the closest answering point.

    b. Not all networks have 9-1-1 capability.

    c. Calls may be blocked by the local answering point facility.

    d. Multiple calls from separate phones are necessary.

9. In most cases, when dialing 9-1-1, the primary public safety answering point (PSAP) is a:

    a. public operator

    b. law enforcement agency

    c. local fire department

    d. county dispatcher

10. Under what circumstances will a primary PSAP transfer a call to a secondary PSAP?

    a. When the call volume of the primary PSAP is overwhelming

    b. When the incident does not require the presence of the primary PSAP

    c. When the incident does not warrant an emergency response

    d. When legal protocols warrant the transfer

11. When civilians walk into the firehouse to report an emergency, firefighters must ensure that vital information must be obtained from the person relaying the event, and this information must be:
    a. reported to the company officer for consideration
    b. documented immediately on a call sheet
    c. transferred to the communications center as rapidly and efficiently as possible
    d. verified with a secondary party for validity

12. This means of communication is still the most common method of reporting calls for assistance:
    a. direct-dial phone
    b. conventional phone
    c. text phones
    d. direct line

13. PSAP is used to describe any center designated for receipt of 9-1-1 calls. What does PSAP stand for?
    a. public safety answering point
    b. public safety affirmation procedure
    c. proactive safety alarm procedure
    d. personnel satisfaction alarm pass

14. Using street boxes as alerting systems to the fire department is problematic because:
    a. They are difficult to see at night.
    b. They are prone to weather-related malfunctions.
    c. They are not available at all locations.
    d. They are prone to false alarms.

15. These types of alarm systems put the caller and the telecommunicator in direct voice contact and are usually recorded.
    a. property monitoring systems
    b. text display boxes
    c. call boxes
    d. redundant systems

16. Calls for incidents such as helping an individual off a floor are nonemergency in nature and are often referred to as:
    a. service calls
    b. nuisance alarms
    c. secondary alarms
    d. still alarms

17. When receiving a call from a civilian requesting assistance, the telecommunicator must:
    a. Take charge of the information flow.
    b. Allow the caller to talk and listen attentively.
    c. Keep the caller on the phone until the conclusion of the incident.
    d. Conference in the emergency response personnel who will be responding.

18. Which of the following is not included in the critical information that must be gathered by the telecommunicator when taking a call for assistance?
    a. contact phone number
    b. location
    c. caller's name
    d. resources required

19. At the emergency scene, how are most communications transmitted to dispatch?
    a. by cell phone
    b. by mobile or portable radios
    c. by computer terminals and networks
    d. by hardwire telephone

20. Suppose your engine is responding to a reported building fire and you witness a vehicle accident occurring at an intersection. What will be the first action to take?
    a. Continue the response to the building fire.
    b. Determine what resources are needed to be dispatched to the accident scene.
    c. Contact the dispatch center and advise the dispatcher of the situation.
    d. Decide whether the engine company should remain on the scene of the vehicle accident or proceed on their original call.

21. The highest proportion of daily communication activity across the radio airways is made up of:
    a. routine message transmissions
    b. fire traffic transmissions
    c. emergency traffic transmissions
    d. Mayday traffic transmissions

22. What is the first step in the communication process?
    a. dispatch
    b. arrival information of the first company
    c. receipt of a request for service
    d. availability of a unit

23. The only time this callout should be used is in a situation in which there is an immediate threat to the lives of firefighters:
    a. emergency traffic
    b. Mayday
    c. urgent
    d. clear the air

24. This type callout in used when a structural problem is discovered, indicating the danger of collapse:
    a. emergency traffic
    b. Mayday
    c. urgent
    d. clear the air

25. A Level 1 personnel accountability system (PAS) takes place:
    a. at shift change
    b. at all working fires
    c. in extreme situations
    d. at the order of the incident commander

26. A Level 2 personnel accountability system (PAS) takes place:
    a. at shift change
    b. at all working fires
    c. in extreme situations
    d. at the order of the incident commander

27. A Level 3 personnel accountability system (PAS) takes place:
    a. at shift change
    b. at all working fires
    c. in extreme situations
    d. at the order of the incident commander

28. Upon reporting for duty, firefighters should take one of their identification name tags and do what with it?
    a. Clip it on their helmet.
    b. Give it to their company officer.
    c. Place it on a status board on the apparatus.
    d. Place it on a status board in the fire station.

29. This is a polling system used by the incident commander or accountability officer to ascertain that all personnel operating at an emergency incident are safe and accounted for.
    a. personnel tracking report
    b. personnel accountability record
    c. personnel accountability system
    d. accountability officer log

30. If during the accounting process, if an individual or company is not responding, what is the next step?
    a. Emergency traffic procedures will be implemented.
    b. An emergency traffic or Mayday will be requested.
    c. An accountability officer will be designated.
    d. Departmental operational strategy and tactics for the missing firefighters will be implemented.

31. The important thing to understand regarding emergency traffic communications used by fire departments is:
    a. It must be thoroughly understood by all personnel.
    b. Training must be conducted on a yearly basis.
    c. Only the incident commander can initiate emergency traffic communications.
    d. Its use is limited to fire incidents only.

32. In regard to radio operations, this agency is the assigning agency and licenses fire departments on one or multiple specific frequencies:
    a. National Fire Protection Association (NFPA)
    b. National Radio Association (NRA)
    c. Federal Communications Commission (FCC)
    d. Occupational Safety and Health Administration (OSHA)

33. These radio systems receive input on one frequency and provide output on a separate frequency and can be used throughout the radio system to provide a stronger signal:
    a. repeater systems
    b. simplex systems
    c. complex systems
    d. direct systems

34. These radio systems provide a better use of the radio frequencies and channels available because they constantly readjust the frequencies used for conversations. This allows for a more efficient use of the limited frequencies because each conversation does not require a dedicated channel.

    a. simplex systems
    b. trunked systems
    c. double-trunked systems
    d. control channel system

35. Fire departments use three different radios for service-related activities. Which of the following is not one of these radios?

    a. portable radio
    b. mobile radio
    c. headset radio
    d. base station

36. A computer-like device that can provide communication and transmit information such as company status, emergency notification, and event information is known as what?

    a. mobile communication system (MCS)
    b. emergency data retriever (EDR)
    c. mobile computer terminal (MCT)
    d. mobile data terminal (MDT)

37. Often communication facilities such as police departments, fire departments, and EMS communications centers use this form of communication to expedite contact with each other:

    a. LAN-based telephones
    b. cellular telephones
    c. direct-connect telephones
    d. personal digital assistants

38. What is a disadvantage of using a numeric code such as a "10-code"?

    a. It uses less air time.
    b. It frees up radio channels for other transmissions.
    c. The transmission is brief.
    d. The transmission can have multiple meanings.

39. What is the primary purpose of the public fire service communication system?

   a. to track companies for the purposes of dispatching while on the road
   b. to accept reports of emergencies and dispatch the appropriate units
   c. to interact with the public on all emergency matters and refer them to the appropriate authority
   d. to provide a mechanism for the issuance of a Mayday transmission

40. From the standpoint of the firefighter assigned to a company, perhaps the most fundamental aspect of fire service communications is:

   a. notifying appropriate emergency services when requested
   b. providing a means for the recording of incident communications
   c. initiating and receiving alarms
   d. receiving and acknowledging transmissions from the fireground

41. In some areas, frequent false alarms led to replacement of some boxes with_____ systems:

   a. voice-capable
   b. encoded
   c. decoded
   d. telegraph

42. For many departments, this is the most common type of fire-related incident:

   a. vehicle fire
   b. rubbish fire
   c. carbon monoxide alarm
   d. automatic fire alarm

43. When handling emergencies reported directly to personnel at a fire station or on apparatus, which of the following information is not required and does need to be transferred to dispatch?

   a. address
   b. type of apparatus to be dispatched
   c. nature of incident
   d. contact information from the reporting party

44. In most communities, this type of card was used to describe both the physical location of the alarm box and the apparatus or resources that were assigned to respond.

   a. alarm card
   b. assignment card
   c. box card
   d. job card

45. The number of units sent on an alarm is a function of this type of policy:

   a. local department
   b. state-mandated
   c. NFPA-required
   d. personnel available

46. The most common means for the alerting of fire service resources is via _____.

   a. pager
   b. radio
   c. telephone
   d. public address

47. This type of radio operation means that one frequency is used for transmitting, while a second frequency is used for receiving:

   a. simplex
   b. dual
   c. duplex
   d. multiversal

48. A capability included in radio systems that displays on the dispatcher's console and allows the dispatcher to know what unit is calling is known as a:

    a. unit identifier
    b. unit referencer
    c. identification triangulation capability
    d. personnel accountability enhancer

49. These radio systems offer the ability to better utilize a limited number of frequencies by aggregating a large pool of users with the anticipation that peak demand for all user groups will occur at different times:

    a. ordinary
    b. conventional
    c. trunked
    d. enhanced

50. In this type of radio system, channels are selected by the user, and switching of frequencies is not usually necessary to complete a conversation with another user:

    a. ordinary
    b. conventional
    c. trunked
    d. enhanced

51. Regarding digital signal technology, which of the following is not true?

    a. Digital signals have the capability for better transmission range.
    b. Digital signals are less susceptible to noise being introduced during the transmission process.
    c. Digital radio systems have a small time delay between the transmitting party's voice and the receiving party.
    d. The digital radio signal being sent between parties on the system is the user's voice sent over the airwaves.

52. This type of radio technology is used to receive transmissions from field units and rebroadcast them at higher power on the receive frequency of user equipment

    a. rebroadcaster
    b. repeater
    c. multicaster
    d. simulcaster

53. Which of the following is extremely valuable in dispatching because it enhances selection of the closest unit to alarms and manages system-wide deployment in real time?

    a. mobile data terminal (MDT)
    b. mobile computer terminal (MCT)
    c. automatic vehicle location (AVL)
    d. enhanced geographic dispatching (EGD)

54. What is the National Incident Management System (NIMS) stance on the use of 10 codes for communication?

    a. They are acceptable except at large incidents.
    b. They are acceptable when adopted by entire mutual aid groups.
    c. They must meet the criteria set forth by the NIMS Integration Center.
    d. They will compromise communications at an incident with multiple jurisdictions.

CHAPTER 4 ANSWER KEY

Question #	Answer	Page #	Question #	Answer	Page #
1	b	49	33	a	55
2	b	58	34	b	55
3	a	59	35	c	61
4	c	60	36	d	57
5	d	59	37	c	50
6	c	51	38	d	69
7	a	51	39	b	50
8	a	52	40	c	50
9	b	52	41	a	50
10	b	51	42	d	52
11	c	51	43	b	53
12	b	51	44	c	60
13	a	58	45	a	60
14	d	52	46	b	60
15	c	50	47	c	54
16	d	50	48	a	54
17	a	53	49	c	54
18	d	61	50	b	54
19	b	61	51	d	55
20	c	62	52	b	55
21	a	62	53	c	57
22	c	63	54	d	63
23	b	64			
24	c	64			
25	a	65			
26	b	65			
27	c	65			
28	c	65			
29	b	65			
30	b	65			
31	a	65–66			
32	c	53			

Fire Behavior

by Sean Gray

1. A fire is best described as:
   a. a chemical reaction that produces flames and smoke
   b. a chemical process that produces flame and particulate matter
   c. a process of oxidation that produces heat and particulate matter
   d. a rapid oxidation process with the evolution of light and heat in varying intensities

2. Sublimation is best defined as:
   a. process of a matter changing from a solid to a gas
   b. process of a liquid changing to a solid
   c. process of a solid changing to a liquid
   d. process of the evaporating of a gas

3. A three-dimensional diagram that illustrates the interrelationship of the chemical chain reaction in the combustion process is known as the:
   a. fire triangle
   b. fire tetrahedron
   c. fire octagon
   d. fire pentagon

4. The heat release rate of burning products is important to know for firefighters because:
   a. It helps determine the correct volume of extinguishing agent to combat the fire.
   b. It dictates how much personal protective equipment is required.
   c. It can determine how long the fire will burn.
   d. It can accurately determine the speed of fire spread.

5. The term *specific heat* can be described as:
   a. amount of heat needed to cause an object to ignite
   b. amount of thermal energy required to raise unit mass of a substance by 1° (its units are J/kg-K)
   c. amount of heat a substance absorbs as the temperature of the substance increases
   d. both B and C

6. The term *latent heat* is best described as:
   a. the amount of thermal energy remaining in a burning object after it is extinguished
   b. the thermal energy expelled when a substance is converted from a solid to a liquid
   c. the thermal energy that is absorbed when a substance is converted from a solid to a liquid, or from a liquid to a gas
   d. the thermal energy expelled when a substance is converted from a liquid to a gas

7. An object with a high-density ratio will need:
   a. an equal amount of heat energy to cause ignition
   b. less heat energy if applied equally across the object to cause ignition
   c. more heat energy to cause ignition
   d. less heat energy to cause ignition

8. If a gas has a higher specific gravity than 1.0, it will be one of the following:
   a. heavier than air
   b. lighter than air
   c. equal to air
   d. equal to or lighter than air

9. What is the amount of heat absorbed by a liquid that passes to a gaseous form called?
   a. high heat
   b. latent heat of vaporization
   c. specific heat
   d. low heat

10. Accepted sources of heat include:
    a. conducted, radiated, and convected
    b. solar, radiated, and chemical
    c. mechanical, radiated, and electrical
    d. chemical, electrical, mechanical, and nuclear

11. The definition "chemical decomposition of a compound into one or more other substances by heat alone" best describes:
    a. vaporization
    b. pyrolysis
    c. open burning
    d. pyrolytic Compounding

12. Why is it possible for certain flammable gases, when escaping under pressure, to ignite with no outside ignition source?
    a. The application of pressure alone can ignite unstable gases.
    b. The oxygen moving through the gas can be heated, causing ignition.
    c. The flowing gas can create its own static electric discharge.
    d. The rapid release of gas can cause it to overheat, causing ignition.

13. A compressed gas cylinder may heat up if it being refilled at too fast a pace. In this case, the heating is an example of a:
    a. mechanical heat source
    b. chemical heat source
    c. electrical heat source
    d. radiated heat source

14. Items that can burn are said to usually be:
    a. nonorganic
    b. mostly inorganic
    c. inorganic
    d. organic

15. One of the characteristics of fuel is that it can be in:
    a. only one state at a time
    b. any of three physical states
    c. any of four physical states
    d. ignitable vapor, but not smoke

16. For a solid substance to ignite and burn, the substance must:
    a. produce its own heat
    b. go through a transformation into a gaseous state
    c. be able to continually absorb oxygen
    d. have an equal surface to mass ratio

17. The greater an object's surface to mass ratio is:
    a. The easier it is ignite.
    b. The harder it is to ignite.
    c. The harder it is to extinguish.
    d. The easier it is to extinguish.

18. The minimum temperature at which an ignitable liquid gives off enough vapor to form an ignitable mixture in air near the surface of the liquid, is known as the:
    a. fuel combustion point
    b. fuel ignition limit
    c. flash point
    d. liquid-ignition limit

19. An ignitable liquid is said to have reached its _____ when it is heated enough to give off sufficient vapors to ignite and sustain burning.
    a. liquid-ignition limit
    b. fuel combustion point
    c. flash point
    d. fire Point

20. At sea level, the percentage of oxygen in air is approximately:
    a. 18.5%
    b. 19.8%
    c. 20.5%
    d. 20.95%

21. The flammable or explosive limits of gaseous fuel have an upper point and a lower point, which represent:
    a. the amount of air mixed with fuel
    b. the amount of oxygen present in the mix
    c. the percentage of fuel vapor in the air
    d. the percentage of oxygen in the vapor

22. A danger faced by firefighters when venting a house where a rich flammable mixture is present is one of the following:
    a. The rich mixture can easily ignite, causing a fire.
    b. The rich mixture can easily ignite, causing an explosion.
    c. The venting of the rich mixture will bring the mixture into an explosive range.
    d. The venting will make the mixture lean, increasing its ability to ignite.

23. The four stages of fire development are:
    a. ignition, rollover, flashover, and fully developed
    b. ignition, growth, fully developed, and decay
    c. ignition, rollover, flashover, and decay
    d. ignition, growth, fully developed, and extinguished

24. For a fuel to ignite, it must produce
    a. ignitable vapors
    b. oxygen
    c. gaseous fumes
    d. heat

25. The instant when a heat source unites with an ignitable vapor in an oxygen sufficient environment that results in a chemical chain reaction is known as:
    a. flash point
    b. fire point
    c. ignition
    d. flame

26. A fire will not be able to remain burning if the available oxygen level falls below approximately:
    a. 20.8%
    b. 8.0%
    c. 16.5%
    d. 16.0%

27. Why will a fire in a well-insulated house develop faster than in a poorly insulated house?
    a. The insulated walls will not absorb any heat, allowing all of the heat to remain in the room.
    b. The windows will fail faster, due to the insulation keeping more heat in the room, and thereby allowing an uninterrupted supply of air.
    c. The insulation in walls will burn, causing the fire to spread faster.
    d. More heat will be reradiated back into the fire room than will pass through the walls, helping to heat furnishings.

28. The phenomenon where hotter and cooler gases stratify in the fire compartment is known as:
    a. thermal layering
    b. thermal balance
    c. heat energy stratification
    d. temperature balancing

29. The ignition of heated gases and smoke at the ceiling level, causing flames to appear, is known as:
    a. black fire
    b. burn-over
    c. rollover
    d. flashover

30. The point where all the contents in a room are giving off ignitable vapors and reach their ignition temperature is known as:
    a. rollover
    b. flashover
    c. burn-over
    d. free-burning

31. An explosive force caused by the reintroduction of oxygen into an oxygen-starved fire compartment is known as:
    a. black fire
    b. backdraft
    c. flashover
    d. oxy-explosive phenomenon

32. Transfer of heat through a solid object is known as:
    a. radiation
    b. convection
    c. conduction
    d. direct contact

33. Transfer of heat through air movement is known as:
    a. convection
    b. radiation
    c. thermal
    d. conduction

34. Transfer of heat through space by light waves is known as:
    a. thermal
    b. radiation
    c. convection
    d. conduction

CHAPTER 5 ANSWER KEY

Question #	Answer	Page #	Question #	Answer	Page #
1	d	70	31	b	83
2	a	71	32	c	77
3	b	70	33	a	77
4	a	74	34	b	78
5	d	73			
6	c	73			
7	c	75			
8	a	73			
9	b	73			
10	d	75			
11	b	77			
12	c	76			
13	a	76			
14	d	71			
15	b	71			
16	b	71			
17	a	71			
18	c	71–72			
19	d	72			
20	d	72			
21	c	72			
22	c	72			
23	b	80			
24	a	80–81			
25	c	80–81			
26	b	81			
27	d	81			
28	a	81			
29	c	82			
30	b	82			

Fire Extinguishers

by W. Jack Miller

1. Fires are classified according to their:
   a. products of combustion
   b. heat release
   c. size
   d. fuel

2. To be effective and safe, the extinguisher being used must match the _____ to be extinguished.
   a. size of the fire
   b. product
   c. class of fire
   d. rate of heat release

3. Combustible cooking media is one type of fire where a _____ extinguisher should not be used.
   a. water
   b. dry chemical
   c. carbon dioxide
   d. wet chemical

4. A short circuit could occur if a stream of water is discharged onto:
   a. de-energized electrical equipment
   b. live electrical equipment
   c. burning ordinary combustibles
   d. both a and b

5. _____ rate fire extinguishers according to the type of fires they can safely extinguish.
   a. Fire prevention bureaus
   b. Fire extinguisher manufacturers
   c. Independent testing laboratories
   d. The National Fire Protection Association

6. Listed extinguishers have been:
   a. tested and approved
   b. produced and tested
   c. inspected and used
   d. tested and inspected

7. Fires involving live electrical equipment can best be extinguished using a _____ fire extinguisher.
   a. Class K
   b. Class C
   c. Class B
   d. Class D

8. Fires involving vegetable oil or animal fats can best be extinguished using a _____ fire extinguisher.
   a. Class D
   b. Class B
   c. Class A
   d. Class K

9. Fires involving combustible metal alloys can best be extinguished using a _____ fire extinguisher.
   a. Class C
   b. Class B
   c. Class D
   d. Class K

10. Fires involving rubber and fabrics can best be extinguished using a _____ fire extinguisher.
    a. Class B
    b. Class A
    c. Class D
    d. Class C

11. Fires involving oils, paint, and lacquer can best be extinguished using a _____ fire extinguisher.
    a. Class C
    b. Class K
    c. Class A
    d. Class B

12. Which class of fire includes natural fibers and may be extinguished using water?
    a. Class K
    b. Class A
    c. Class B
    d. Class C

13. Which class of fire includes magnesium and zirconium and will require specifically designed chemical extinguishing agents to put out?
    a. Class B
    b. Class C
    c. Class K
    d. Class D

14. Which class of fire involves energized electrical equipment and can be extinguished by using carbon dioxide?
    a. Class D
    b. Class C
    c. Class A
    d. Class B

15. All portable fire extinguishers use _____ to expel the extinguishing agent.
    a. pressure
    b. gravity
    c. chemical reaction
    d. heat

16. Some fire extinguishers store the pressurizing gas externally using a:
    a. pressurizing tube
    b. separate cylinder
    c. gas pressurized collar
    d. pressurizing link

17. Carbon dioxide is an example of an extinguishing agent that:
    a. must be pressurized externally
    b. cannot be pressurized
    c. creates its own pressure
    d. requires the addition of a pressurizing agent

18. The majority of fire extinguishers consist of the following six basic parts: cylinder, handle, lever, nozzle or horn, pressure indicator, and:
    a. inspection tag
    b. base plate
    c. locking mechanism
    d. fill port

19. Which type of extinguisher keeps the agent and the pressurizing gas together in the cylinder?
    a. cartridge and cylinder
    b. stored cylinder
    c. pressurized cylinder
    d. stored pressure

20. To discharge the extinguishing agent, which part of handle is depressed?
    a. handle guard
    b. lever
    c. handle lock
    d. horn

21. The purpose of the _____ is to prevent accidental discharge.
    a. locking mechanism
    b. handle lock
    c. nozzle guard
    d. handle guard

22. As an extinguishing agent, water is best suited for _____ fires.
    a. Class K
    b. Class A
    c. Class C
    d. Class D

23. Water extinguishes fires by:
    a. interrupting the fire's chemical bonding
    b. cooling fuel to below its kindling temperature
    c. wetting exposed material
    d. limiting the available oxygen

24. An extinguishing agent that creates bubbles that form a film when breaking, creating a film on the surface of the fuel to prevent the escape of vapors, best describes:
    a. fluoroprotein foam
    b. alcohol foam
    c. aqueous film-forming foam
    d. carbon dioxide

25. An extinguishing agent that works best against fires involving polar solvents is:
    a. water
    b. film-forming fluoroprotein foam
    c. aqueous film-forming foam
    d. alcohol type aqueous film-forming foam

26. Film-forming fluoroprotein (FFFP) foam is similar to aqueous film-forming foam (AFFF) in that it:
    a. will not allow a film to form on a flammable liquid.
    b. allows a film to form under the surface of a burning liquid.
    c. separates the fuel from the heat source.
    d. forms a film over a flammable liquid surface.

27. Wet chemical agents are specifically designed for which class of fire?
    a. Class B
    b. Class C
    c. Class D
    d. Class K

28. When using wet chemical, the reaction in which a soapy foam forms on top of the burning material of the cooking medium is known as:
    a. foam residue formation
    b. saponification
    c. aqueous film formation
    d. surface reaction

29. To be successfully extinguished, cooking medium fires must be:
    a. cooled below their ignition temperature
    b. cooled to a point equal to their ignition temperature
    c. continually checked for reignition
    d. fought with dry chemical extinguishing agents

30. This type of extinguisher discharges a fine spray and is designed as an alternative to halon:
    a. foam-mist extinguisher
    b. water-mist extinguisher
    c. Bresnan distributor
    d. aspirated dry chemical distributor

31. The water-mist extinguisher may be used on Class C fires.
    a. true
    b. false

32. Which class of fires requires the process of saponification to occur in order to extinguish a fire?
    a. Class K
    b. Class B
    c. Class D
    d. Class C

33. Which class of fires includes ignitable lubricants and may be extinguished using dry chemical extinguishing agents?
    a. Class D
    b. Class K
    c. Class B
    d. Class C

34. Purple K, ABC, Super K, and Monnex are all examples of this type of extinguishing agent:
    a. wet chemical
    b. foaming agents
    c. dry chemical
    d. halogenated

35. Why is dry chemical not recommended for fires in delicate electrical equipment?
    a. It will not completely extinguish the fire.
    b. The dry chemical will conduct electricity.
    c. The dry chemical is an irritant.
    d. It may damage the equipment beyond repair.

36. Sodium bicarbonate is also known as:
    a. regular dry chemical
    b. BC
    c. wet chemical
    d. both A and B

37. Why are regular dry chemical extinguishers not recommended for Class A fires?
    a. They have no effect on deep-seated fires.
    b. The nozzle design does not allow for efficient agent application.
    c. They lack sufficient discharge pressure.
    d. The extinguishing agent lacks sufficient moisture content.

38. Purple K uses a _____ extinguishing agent:
    a. potassium bicarbonate
    b. potassium chloride
    c. sodium bicarbonate
    d. urea potassium

39. Monoammonium phosphate dry chemical is more commonly known as:
    a. BC
    b. ABC
    c. Super K
    d. Monnex

40. ABC dry chemical extinguishes ordinary combustible fires by creating a _____ on the surface of the burning material.
    a. molten residue
    b. chemical barrier
    c. mild chars
    d. cooling residue

41. Burning lithium is an example of a Class _____ fire.
    a. A
    b. B
    c. C
    d. D

42. Class D agents work by excluding oxygen and by :
    a. creating a thermal barrier
    b. allowing thermal energy to pass through the agent
    c. absorbing the thermal energy
    d. halting the fire's chemical chain reaction

43. The Class D extinguishing agent graphite is best applied using a:
    a. pressurized fire extinguisher
    b. dry scoop or shovel
    c. hoseline
    d. fire extinguisher and applicator wand

44. Halogenated extinguishing agents suppress fires by interrupting the _____ of the fire.
    a. chemical chain reaction
    b. thermal discharge
    c. production of by-products
    d. oxygen supply

45. The primary advantage of halogenated agents has been:
    a. environmental safety
    b. lack of cleanup required
    c. low replacement cost
    d. application options

46. FE-36 and Halotron are examples of:
    a. halogenated agents
    b. new wet chemical agents
    c. hose-applied dry chemical agents
    d. extinguisher applied halon alternatives

47. FE-36 and Halotron extinguish fires primarily through:
    a. random chain breakage
    b. cooling
    c. oxygen depletion
    d. wetting

48. Because a violent reaction could occur, these types of extinguishers should not be used on fires involving oxidizers:
    a. ABC dry chemical, Halon 1211, and Halotron I
    b. Halon 1211 and Halotron I
    c. ABC dry chemical and Halotron I
    d. Halon 1211 and ABC dry chemical

49. During Class A fire tests, the minimum allowable discharge time for extinguishers rated 2A and higher is:
    a. 10 seconds
    b. 13 seconds
    c. 30 seconds
    d. 1 minute

50. For an extinguisher to be classified for Class C fires, the extinguisher test consists of:
    a. the extinguishing agent rapidly controlling a live electrical fire
    b. the extinguishing agent extinguishing a live electrical fire in less than 1 minute
    c. the extinguishing agent only needing to be tested for electrical nonconductivity
    d. the extinguishing agent extinguishing a Class C fire in less than 30 seconds

51. The minimum rate at which the agent must be applied to the fire, usually expressed in a ratio of quantity of agent per area unit per time unit, is called the:
    a. critical agent rate
    b. agent application rating
    c. minimum application rate
    d. critical application rate

52. The disadvantages of using higher flow rates include:
    a. wasting chemicals
    b. having less discharge time
    c. less protection from radiant heat
    d. both a and b

53. Theoretically, two operators applying dry chemical at the same time can _____ the flow rate of their extinguishers.
    a. double
    b. triple
    c. quadruple
    d. The flow rate remains the same.

54. When operating a fire extinguisher, the acronym PASS stands for:
    a. pull, arm, squeeze, sweep
    b. position, aim, squeeze, sweep
    c. position, arm, squeeze, sweep
    d. pull, aim, squeeze, sweep

55. The most important aspect of any fire extinguisher application technique is:
    a. application time
    b. operator safety
    c. amount of agent applied
    d. choice of extinguishing agent

56. With regard to protection against radiant heat, foam streams:
    a. give good protection
    b. do not give good protection
    c. will only initially give good protection
    d. should only be used after the main body of fire is controlled

57. When applying foam to a fire, it should be applied:
    a. gently with minimal disturbance to the fuel surface
    b. aggressively to quickly knock down the fire
    c. below the fuel surface
    d. from as close to the burning fuel surface as possible

58. If water-miscible fuels are involved in the fire, a/an _____ must be used.
    a. film-forming foam
    b. hydrocarbon type foam
    c. alcohol type foam
    d. dry chemical

59. Because dry chemical uses a _____ action to extinguish a Class B fire, no physical properties of the fire are affected.
    a. wetting
    b. smothering
    c. flooding
    d. chain-breaking

60. Flammable liquid contained in a certain area by a vessel, berm, or dike is considered to be:
    a. fuel in depth
    b. puddle fuel
    c. deep fuel
    d. controlled depth fuel

61. When fighting in-depth flammable liquid fires with dry chemical extinguishers, the sweep of the extinguisher should be _____ inches past the edge of the burning fuel.
    a. 12
    b. 10
    c. 8
    d. 6

62. Three-dimensional Class B fires involve flammable liquids:
    a. in motion
    b. in containment
    c. under pressure
    d. in depth

63. When attacking a three dimensional flammable liquids fire, the point of attack should begin with the:
    a. source of the flow
    b. ground fire
    c. exposed objects or properties
    d. valve control area

64. One of the most successful ways of combating fires with extinguishers is by:
    a. working independently
    b. working consecutively
    c. working in teams
    d. working individually

65. For fires involving flammable liquids under pressure, fast-flow equipment with high flow rates must be used because:
    a. The water application flow rate must match the escaping fuel flow rate.
    b. The amount of extinguishing agent needed could clog a normal nozzle.
    c. That is the only mechanism to apply the dry chemical agent.
    d. The dry chemical flow rate must match the escaping fuel flow rate.

66. When used in conjunction with water hoselines on fog patterns, dry chemical can be applied to fires with:
    a. little success
    b. great success
    c. moderate success
    d. no success

67. When applying dry powder to Class D fires, the operator should be careful that _____ is present in the surrounding area.
    a. no moisture
    b. no ignition source
    c. no additional fuel
    d. no wind condition

68. Fire extinguishers shall be inspected at a minimum of:
    a. 7-day intervals
    b. 30-day intervals
    c. 6-month intervals
    d. 1-year intervals

69. The test in which the extinguisher cylinder is subjected to pressure applied by water or other noncompressible fluid is known as a _____ test.
    a. water-pressure test
    b. cylinder pressure test
    c. hydrostatic test
    d. cylinder-static test

70. Carbon dioxide extinguishers need to hydrostatically tested every _____ years.
    a. 3
    b. 5
    c. 6
    d. 12

71. Extinguishers requiring 12-year hydrostatic tests shall be emptied and subject to a thorough examination of mechanical parts every _____ years.
    a. 12
    b. 3
    c. 4
    d. 6

72. Stored-pressure water extinguishers with fiberglass shells, vaporizing liquid, and cartridge-operated loaded stream extinguishers, are examples of extinguishers that are:

    a. obsolete

    b. made for civilian use

    c. required to undergo annual hydrostatic testing

    d. expected to remain in use for the fire service for the foreseeable future

CHAPTER 6 ANSWER KEY

Question #	Answer	Page #	Question #	Answer	Page #
1	d	89	33	c	94
2	c	89	34	c	94
3	a	92	35	d	94
4	b	92	36	d	94
5	c	90	37	a	94
6	a	90–91	38	a	94
7	b	90–91	39	b	94
8	d	90–91	40	a	94
9	c	90–91	41	d	94
10	b	90–91	42	c	94
11	d	90–91	43	b	94
12	b	90–91	44	a	95
13	d	94	45	b	94–95
14	b	93	46	d	95
15	a	92	47	b	95
16	b	92	48	a	97
17	c	92	49	b	95
18	c	92	50	c	97
19	d	92	51	d	97
20	b	92	52	d	98
21	a	92	53	a	98
22	b	92	54	d	98–99
23	b	92	55	b	99
24	c	92	56	b	99
25	d	92	57	a	99
26	d	92	58	c	100
27	d	93	59	d	100
28	b	93	60	a	100
29	a	93	61	d	100
30	b	93	62	a	100
31	a	93	63	b	100–101
32	a	93	64	c	101

CHAPTER 6 ANSWER KEY *Continued*

Question #	Answer	Page #
65	d	101
66	b	101
67	a	101
68	b	101
69	c	102
70	b	103
71	d	102
72	a	103

Building Construction

by Paul T. Dansbach

1. A load calculated into a building based on sound engineering principles is known as a(n):
   a. fire load
   b. axial load
   c. designed load
   d. dead load

2. Unauthorized construction may allow this type of load to be introduced into an existing building:
   a. undesigned
   b. eccentric
   c. torsional
   d. axial

3. People and transportable items are examples of:
   a. undesigned loads
   b. live loads
   c. designed loads
   d. dead loads

4. Interior finishes and built-in components are examples of:
   a. torsional loads
   b. impact loads
   c. undesigned loads
   d. dead loads

5. Code requirements for these types of loads vary based on geographic areas:
   a. eccentric load
   b. environmental load
   c. fire load
   d. static load

6. This type of load is the opposite of a force applied slowly over a long period of time:
   a. impact load
   b. environmental load
   c. static load
   d. live load

7. A load that is relatively unchanged over an extended period of time is known as a/an:
   a. live load
   b. environmental load
   c. static load
   d. undesigned load

8. A force that passes through the center of a supporting member is called a/an:
   a. torsional load
   b. axial load
   c. eccentric load
   d. center pitched

9. A load passing through a structural member at a point other than the center of that member is known as a/an:
   a. eccentric load
   b. concentric load
   c. off-center pitched load
   d. axial load

10. A load that is parallel to the cross section of the supporting column and does not pass through the long axis of that column is called a/an:
    a. obtuse load
    b. parallel-centered load
    c. concentric load
    d. torsional load

11. This type of load is commonly measured in pounds per square foot:
    a. static load
    b. fire load
    c. designed load
    d. undesigned load

12. In a wood frame building, this type of structural connection is susceptible to early burn-through and collapse:
    a. mortar and tendon
    b. mortar and tenon
    c. mortise and tenon
    d. mortise and tendon

13. The primary structural element in a noncombustible building is:
    a. structural steel
    b. cement block
    c. concrete
    d. dimensional lumber

14. This dense, noncombustible material can be used to make floors, walls, and columns:
    a. cement block
    b. cast-in-place concrete
    c. cast iron
    d. gypsum

15. This material may fracture if exposed to high temperatures:
    a. wood
    b. structural steel
    c. cold-formed steel
    d. cast iron

16. This horizontal building component transfers weight from one structural member to another structural member:
    a. decking
    b. wall
    c. beam
    d. column

17. This building component is supported at only one end:
    a. girder
    b. cantilever beam
    c. truss
    d. joist

18. A structural component consisting of a top and bottom flange, separated by a vertical piece known as the web, is called a/an:
    a. I beam
    b. T beam
    c. I truss
    d. fire cut beam

19. A building's ability to resist collapse is known as its:
    a. structural hierarchy
    b. structural integrity
    c. structural resistance
    d. structural stability

20. For firefighting purposes, buildings are classified in five groups based on their:
    a. fire-suppression and detection systems
    b. life hazard
    c. fire resistiveness
    d. type of occupancy

21. A building of Type I construction is considered to be:
    a. mill or heavy timber construction
    b. fire-resistive construction
    c. wood frame construction
    d. ordinary construction

22. Mill or heavy timber construction is classified as _____ construction:
    a. Type I
    b. Type II
    c. Type III
    d. Type IV

23. A building of Type II construction is considered to be:
    a. noncombustible
    b. combustible
    c. fire resistive
    d. ordinary

24. Wood frame construction is classified as _____ construction:
    a. Type V
    b. Type IV
    c. Type III
    d. Type II

25. A building of Type III construction is considered to be:
    a. ordinary
    b. noncombustible
    c. wood frame
    d. heavy timber

26. Buildings that have columns, beams, and floor slabs that are protected from the heat of a fire are considered to be of _____ construction.
    a. mill
    b. ordinary
    c. noncombustible
    d. fire-resistive

27. To protect structural steel in a fire-resistive building, all of the following can be used except:
    a. gypsum board
    b. spray-on fire proofing material
    c. intumescent paint
    d. cement plaster

28. Fire-resistive construction can be expected to be found in all of the following occupancies except:
    a. hospitals
    b. high-rise buildings
    c. townhouses
    d. cold-storage buildings

29. Most modern high-rise buildings are constructed with:
    a. components heavier in weight than in the first part of the 20th century
    b. components lighter in weight than in the first part of the 20th century
    c. components equal in weight to first part of the 20th century
    d. component weight varies between regions and cannot be considered standard

30. Many new high-rise office buildings are designed using a _____ floor plan:
    a. center core
    b. limited center core
    c. four corner escape
    d. utility core

31. Because of limited room size and fire separation walls to the adjoining corridor, this type of fire-resistive occupancy typically has the least fire spread potential:
    a. high-rise office buildings
    b. high-rise luxury multiple dwellings
    c. mid-rise age-restricted multiple dwellings
    d. high-rise hotels

32. This type of fire-resistive occupancy has the smallest square footage of multiple dwelling buildings, and its greatest threat of fire extension is a dwelling unit door that is left open:
    a. high-rise hotels
    b. high-rise multiple dwellings
    c. high-rise public housing multiple dwellings
    d. high-rise luxury multiple dwellings

33. This type of high-rise occupancy has the greatest potential for fire spread:
    a. high-rise multiple dwellings
    b. high-rise office buildings
    c. high-rise hotels
    d. high-rise public housing multiple dwellings

34. A major difference between fire-resistive construction and noncombustible construction is:
    a. In fire-resistive construction, the structural steel must be protected from the fire's heat.
    b. The overall floor plan design.
    c. The building dimensions.
    d. The type of building materials.

35. A collapse danger of noncombustible constructed buildings is:
    a. spalling of concrete walls
    b. lack of fireproofing material applied to the structural steel
    c. parapet wall collapse
    d. floor separation

36. In a noncombustible constructed building, the exterior walls will be:
    a. bearing walls
    b. walls only at the front and rear of the structure
    c. bearing walls only at the two sides of the structure
    d. a nonbearing wall that supports only itself

37. This type of exterior covering is secured to the steel structure and may be made of various materials, including concrete or mirrored panels:
    a. attachment wall
    b. side curtain
    c. curtain wall
    d. block

38. This type of wall system is attached to the building's steel frame at the roof level and is supported on their other end on the building foundation:
    a. tilt slab wall panels
    b. bearing block wall system
    c. nonbearing block wall system
    d. cast in place concrete walls

39. The collapse zone to be maintained for buildings utilizing tilt slab wall systems should be:
    a. 50% of the height of the wall section
    b. 75% of the height of the wall section
    c. 100% of the height of the wall section
    d. in excess of 100% of the height of the wall section

40. The collapse danger associated with noncombustible buildings is considered:
    a. about the same as other building classifications
    b. low, due to the presence of a large-diameter structural steel framework
    c. low, due to the presence of nonbearing exterior walls
    d. high, due to the presence of bar joist trusses and tilt slab wall panels

41. A structure built with load-bearing masonry walls and wood joist floors and roofs is classified as:
    a. Type II
    b. Type III
    c. Type IV
    d. Type V

42. A fire service term used to describe many ordinary constructed buildings in a downtown commercial district is:

    a. mortgage payer

    b. ratepayer

    c. taxpayer

    d. utility payer

43. Platform, braced, lightweight, and balloon are all types of this classification of construction:

    a. ordinary

    b. Type V

    c. mill

    d. Type I

44. Mortise and tenon connections are used in:

    a. platform construction

    b. post and beam construction

    c. braced frame construction

    d. both b and c

45. Post and beam construction is susceptible to which type of collapse?

    a. inward/outward wall collapse

    b. lean-to floor collapse

    c. pancake floor collapse

    d. lean-over building collapse

46. In this type of wood frame construction, exterior wall studs extend the full height of the building, creating open vertical passages:

    a. braced frame

    b. platform

    c. balloon frame

    d. lightweight

47. A major problem faced by firefighters operating in buildings that have been altered over time is:

    a. changed exit locations

    b. combustible void spaces

    c. nonapproved dwelling units

    d. alterations to gas and electric utilities

CHAPTER 7 ANSWER KEY

Question #	Answer	Page #	Question #	Answer	Page #
1	c	109	31	d	120
2	a	109–110	32	c	120
3	b	110	33	b	120
4	d	110	34	a	119
5	b	110	35	b	117
6	a	110	36	d	120
7	c	110	37	c	120
8	b	109	38	a	121
9	a	109	39	d	121
10	d	109	40	d	121
11	b	110	41	b	118
12	c	125	42	c	122
13	a	119	43	b	118
14	b	118	44	d	125
15	d	112	45	a	125
16	c	111	46	c	126
17	b	111	47	b	128
18	a	111			
19	d	117			
20	c	118			
21	b	118			
22	d	118			
23	a	118			
24	a	118			
25	a	118			
26	d	118			
27	c	118			
28	c	118			
29	b	118			
30	a	119			

Ropes and Knots

by Mark A. Sulcov

1. Rope can be made from _____.
   a. natural material
   b. synthetic material
   c. both synthetic and natural material
   d. any nonbiodegradable material

2. The National Fire Protection Standard developed for life safety rope and equipment is:
   a. NFPA 1403
   b. NFPA 1901
   c. NFPA 1021
   d. NFPA 1983

3. An example of a rope made from natural fibers is _____.
   a. Manila
   b. polyester
   c. nylon
   d. polyethylene

4. Natural ropes suffer from several problems, including mildew, rot, deterioration, and _____.
   a. shrinkage
   b. poor abrasion resistance
   c. poor size to strength ability
   d. both poor abrasion resistance and poor size to strength ability

5. Size to strength ability can be expressed as _____.
   a. the relationship of rope length to rope thickness
   b. the relationship of rope thickness to rope strength
   c. the relationship of rope thickness to deterioration
   d. the relationship of rope diameter to rope length

6. This natural fiber is used make rope of only very small diameters:
   a. sisal
   b. Manila
   c. cotton
   d. nylon

7. This rope, while being soft and pliable, is very low in strength:
   a. cotton
   b. sisal
   c. polyester
   d. Manila

8. This rope is made from the twisting of plant fibers to create longer strands:
   a. cotton
   b. sisal
   c. polyester
   d. Manila

9. The biggest advantage of synthetic material with rope is
   a. Resistance to rot.
   b. Resistance to mildew.
   c. All the rope fibers have the same properties.
   d. Resistance to abrasion.

10. This rope is made from multifilament lines, either twisted into strands or covered with sheathing:
    a. polyester
    b. nylon
    c. polypropylene
    d. polyethylene

11. This kind of rope material, while being resistant to most acids, does not hold up well if shock loaded:
    a. nylon
    b. polyethylene
    c. polyester
    d. polypropylene

12. This kind of rope material is primarily used for water applications:
    a. nylon
    b. polypropylene
    c. polyester
    d. sisal

13. This kind of rope material has a strong resistance to chemicals and acids and is available in many colors:
    a. polyester
    b. nylon
    c. cotton
    d. polyethylene

14. For fire department life safety applications, such as rappelling and hauling, this type of rope preferred:
    a. static
    b. semistatic
    c. semidynamic
    d. dynamic

15. The type of rope construction that typically twists three strands together to make the rope is known as:
    a. braided
    b. braid-twisted
    c. laid
    d. braid-on-braid

16. A major problem of twisted rope is that all the strands are _____.
    a. easily back-twisted
    b. exposed to abuse
    c. twisted to determine rope strength
    d. only made from natural fibers

17. This type of rope construction reduces rope twist when used for rappelling:
    a. laid
    b. braid-on-braid
    c. braided
    d. back-twisted

18. The only braided rope that the fire service will use is constructed of _____ fibers.
    a. synthetic
    b. natural
    c. either natural or synthetic
    d. neither natural nor synthetic

19. Rope constructed with an outer jacket, usually with a distinctive pattern, over a braided rope is known as:
    a. braided
    b. braid-on-braid
    c. laid
    d. kernmantle

20. With braid-on-braid constructed rope, when an inspection discovers sliding of the inner and outer braid, the rope _____.
    a. needs to be inspected for additional damage
    b. has attracted moisture and needs to be dried sufficiently
    c. should be designated as back up life safety rope
    d. must be taken out of service

21. Kernmantle rope is constructed of _____ section(s).
    a. one
    b. two
    c. three
    d. four

22. In kernmantle rope, the outer section is known as the:
    a. outer shell
    b. outer jacket
    c. mantle
    d. kern

23. Kernmantle rope can be made from _____ type(s) of kerns:
    a. one
    b. two
    c. three
    d. four

24. To create static kernmantle rope, which type of kern is used?
    a. laid
    b. twisted
    c. braided
    d. continuous filament fiber

25. When using kernmantle rope, the mantle carries approximately _____ of the load:
    a. 75%–80%
    b. 20%–25%
    c. 45%–50%
    d. 50%–55%

26. This type of rope is used primarily for life rescue rope:
    a. kernmantle
    b. braided
    c. braid-on-braid
    d. laid

27. When inspecting kernmantle rope, the inspector should feel for:
    a. discoloration
    b. moisture
    c. hard and soft spots
    d. even minor amount of picks

28. Rope used for non-life-safety search and rescue operations should be categorized as:
    a. search and rescue rope
    b. life safety rope only
    c. either utility or life safety rope
    d. utility rope only

29. The part of the rope that is used to tie the knot is called the:
    a. running end
    b. working end
    c. standing part
    d. tying end

30. The portion of the rope directly above or below the knot is known as the:
    a. working part
    b. running part
    c. standing part
    d. tied part

31. After the knot is tied, the remaining portion of the rope used for hoisting or pulling is called the:
    a. working end
    b. running end
    c. standing end
    d. pulling end

32. When tying a knot, bringing the rope back along itself side by side is done to create a:
    a. loop
    b. round turn
    c. bend
    d. bight

33. Placing a loop in the rope and having the standing portion of the rope return in the direction it came from creates a:
    a. round turn
    b. double loop
    c. bight
    d. loop

34. When tying a knot, a _____ is made by placing a twist in the rope and having the standing part of rope continue in its initial direction.
    a. hitch
    b. bight
    c. loop
    d. round turn

35. Which of the following is not considered part of the fire service family of knots?
    a. figure eight
    b. figure eight on a bight
    c. double loop figure eight
    d. figure eight follow-through

36. This knot is used so the loose end does not untie:
    a. becket bend
    b. overhand knot
    c. water knot
    d. clove hitch

37. After making a loop in the standing part of the rope, slide the loop over the object being hoisted, making sure the running end passes under the working end, so the rope can tighten up on itself. This best describes how to tie a:
    a. half hitch
    b. becket bend
    c. bowline
    d. water knot

38. Which of the following knots can be tied in either an open or closed manner?
    a. water knot
    b. figure eight follow-through
    c. clove hitch
    d. half hitch

39. Which of the following knots ties together the ends of two ropes?
    a. bowline
    b. figure eight
    c. figure eight on a bight
    d. becket bend

40. This knot is used to form a secure loop under tension:
    a. becket bend
    b. bowline
    c. open clove hitch
    d. closed clove hitch

41. Which of the following knots is used to webbing together?
    a. bowline
    b. water knot
    c. figure eight
    d. becket bend

42. When hoisting equipment, a _____ is used to keep the object away from obstructions:
    a. tie-off line
    b. standing line
    c. running line
    d. tag line

43. When hoisting an axe, the best knot to use adjacent to the tool head is the:
    a. clove hitch
    b. half hitch
    c. bowline
    d. water knot

44. When hoisting a hoseline, the tag line is tied off to the:
    a. Bale of the nozzle.
    b. First coupling.
    c. No tag line is required.
    d. Second coupling.

45. The best knot to use as the lifting knot when hoisting a ladder, is the:
    a. figure eight on a bight
    b. bowline
    c. clove hitch
    d. figure eight follow-through

46. Which of the following methods is not an accepted way to clean rope?
    a. top-loading washing machine
    b. front-loading washing machine
    c. rope washer connected to a sink faucet
    d. brushing with a stiff broom or brush

47. The preferred method for drying wet rope is:
    a. placing it outside on a clean tarp to dry in the sunlight
    b. putting it in a clothes dryer following manufacturer's instructions
    c. laying it flat over a ladder away from direct sunlight
    d. hanging it outside while placed in a mesh bag

48. When placing rope into a dedicated search bag, what is the purpose of tying a figure eight on a bight knot beneath the grommet in the bag?
    a. It stops the rope from falling from the bag.
    b. It will allow for a handhold to pull rope from the bag.
    c. It will allow the bight to be tied off to an exterior area.
    d. Both b and c are correct.

49. When you finish placing rope into a drop bag, you should_____.
    a. Tie a figure eight on a bight knot in the end of the rope for safety.
    b. Tie a figure eight knot in the end of the rope so the end can easily be grasped.
    c. Make sure the rope end is free of knots, as they will need to be untied to use the rope.
    d. Make sure the rope end is pushed into the bottom of the bag so it won't fall out.

50. When you finish coiling a rope, what knot should you use to tie the bight and the end of the rope together?
    a. closed clove hitch
    b. bowline
    c. overhand
    d. becket bend

CHAPTER 8 ANSWER KEY

Question #	Answer	Page #	Question #	Answer	Page #
1	c	138	30	c	141
2	d	138	31	b	141
3	a	138	32	d	141
4	d	138	33	a	141
5	b	138	34	c	141
6	a	138	35	c	141
7	a	138	36	b	142
8	d	138	37	a	142
9	c	139	38	c	142
10	b	139	39	d	142
11	c	139	40	b	143
12	b	139	41	b	144
13	d	139	42	d	144
14	a	139	43	a	146
15	c	139	44	c	146
16	b	140	45	b	146
17	c	140	46	a	149
18	a	140	47	c	150
19	b	140	48	d	150
20	d	140	49	b	150
21	b	140	50	d	142
22	c	140			
23	c	140			
24	d	140			
25	b	140			
26	a	140			
27	c	148			
28	d	141			
29	b	141			

Personal Protective Equipment

by Tim Pillsworth

1. When operating at alarms, your first line of defense is:
   a. your self-contained breathing apparatus (SCBA)
   b. your personal protective equipment
   c. the water discharged from the hoseline
   d. your accountability tag

2. Your personal protective equipment (PPE) is designed to protect you from _____ risks.
   a. day-to-day
   b. extraordinary
   c. dangerous substance
   d. collapse

3. Used in wildland firefighting, this protective item is made of aluminized material to reflect the extreme heat during a fire rollover:
   a. Nomex hood
   b. personal heat shield
   c. personal fire shelter
   d. rollover guard

4. Proximity PPE is used during these type fires:
   a. wildland and aircraft based fires
   b. aircraft based fires only
   c. petroleum based fires only
   d. aircraft and petroleum-based fires

5. How does medical PPE differ from wildland or technical rescue PPE?
   a. Medical PPE protects against needle sticks.
   b. Medical PPE protects the wearer from bloodborne pathogens.
   c. Medical PPE is heavier and less fire resistant.
   d. Medical PPE is more durable.

6. This is the most technical form of PPE:
    a. technical rescue PPE
    b. structural firefighting PPE
    c. hazardous materials PPE
    d. medical PPE

7. This level of hazardous materials protection is designed to protect the wearer from liquids and requires an SCBA to be donned:
    a. Level A
    b. Level B
    c. Level C
    d. Level D

8. This level of hazardous materials protection would be covered by your issued structural PPE:
    a. Level A
    b. Level B
    c. Level C
    d. Level D

9. This level of hazardous materials protection is designed to protect the wearer from liquids or vapors and requires full SCBA:
    a. Level A
    b. Level B
    c. Level C
    d. Level D

10. This level of hazardous materials protection protects the wearer from liquids and does not require the use of SCBA:
    a. Level A
    b. Level B
    c. Level C
    d. Level D

11. The acronym PASS stands for:
    a. personal armed security system
    b. personnel area safety system
    c. personal alert safety system
    d. personnel alarm safety system

12. How would you manually activate your PASS device?
    a. Use your portable radio.
    b. Use voice activation.
    c. Push a button or turn a switch.
    d. Remain still for 30 seconds.

13. In regard to firefighting, you are not protected unless:
    a. Your SCBA is properly worn.
    b. You are properly trained.
    c. Your PASS unit is armed.
    d. All parts of your PPE are worn correctly.

14. What part of your protective structural ensemble is donned first?
    a. turnout pants
    b. boots
    c. helmet
    d. pants

15. Which type of boot offers the greatest protection from water, is the least expensive to purchase, is easily donned, but offers the lowest level of foot and ankle support?
    a. rubber bunker boots
    b. day boots
    c. leather bunker boots
    d. high-tech material boots

16. Of the three pants and coat layers, which is most durable?
    a. outer shell
    b. moisture barrier
    c. thermal liner
    d. thermal barrier

17. Which of the following is an indicator of a heat event on the turnout coat?
    a. shredding
    b. discoloration
    c. melting
    d. annealing

18. This is the thinnest layer of the protective coat or pants:
    a. outer shell
    b. moisture barrier
    c. thermal liner
    d. thermal barrier

19. This device, built into the coat, offers a safe and efficient rescue method for a downed firefighter
    a. drag rescue device (DRD)
    b. rescue assist device (RAD)
    c. rescue web device (RWD)
    d. pull harness device (PHD)

20. When should the helmet chin strap be in place?
    a. whenever the SCBA facepiece is donned
    b. at all times, under all circumstances
    c. only when operating at elevated positions
    d. only when in the interior of a structure

21. You are using a tool on the fireground and you do not have an SCBA facepiece in place. In this case, what other eye protection must be worn?
    a. The Bourke shield from your helmet must be in place.
    b. The pull-down face shield must be in place.
    c. Goggles must be worn.
    d. No other eye protection is required.

22. Once learned, how long should to take to correctly don PPE?
    a. under 1 minute
    b. 90 seconds
    c. 2 minutes
    d. 3 minutes

23. You are about to doff superheated gear. What is the first action you should take?
    a. Keep your gloves on and remove your regulator from your facepiece.
    b. Apply water to the gear to cool it.
    c. Wait until the gear cools before doffing it.
    d. Open the gear from top down to release trapped heat.

24. Because they are life safety tools, which of the following should be placed in an easily accessible area when full PPE is donned?
    a. shove knife and screwdriver
    b. spanner wrench and knife
    c. knife and wire cutters
    d. webbing and medical gloves

25. Where is the best place to keep the abovementioned life safety tools?
    a. in the inside coat pocket
    b. in an outside coat pocket
    c. inside the brim of the helmet
    d. inside the pants pocket

26. According to the NFPA standard, how often should PPE be inspected, cleaned, and repaired (if required)?
    a. monthly
    b. yearly
    c. twice yearly
    d. every 2 years

27. As a line firefighter, how often should you personally inspect your PPE?
    a. at the beginning of every tour of duty
    b. after every alarm
    c. at the end of every tour of duty
    d. on a monthly basis

28. When damage to PPE is found, what action should be taken?
    a. The gear should be tested.
    b. The gear should be only used for training purposes.
    c. The gear should be taken out of service.
    d. The gear should be washed first to better determine the extent of the damage.

29. Why is it not recommended that your turnout gear be brought home for washing?
    a. It is property of the fire department and should not be removed from quarters without permission.
    b. It might be needed to answer an alarm while it is at home.
    c. Home washing machines are not designed to clean firefighting equipment.
    d. The gear may contaminate the home washing machine and the home.

30. When cleaning your gear, which part should be washed first?
    a. outer shell
    b. moisture barrier
    c. thermal liner
    d. thermal barrier

CHAPTER 9 ANSWER KEY

Question #	Answer	Page #
1	b	455
2	a	155
3	c	157
4	d	157
5	b	158
6	c	159
7	b	159
8	d	159
9	a	159
10	c	159
11	c	161
12	c	161
13	d	161
14	b	162
15	a	162
16	a	163
17	b	163
18	b	163
19	a	167
20	b	167
21	c	168
22	a	168
23	d	170
24	c	171
25	d	171
26	c	171
27	b	172
28	c	172
29	d	172
30	c	172

Self-Contained Breathing Apparatus

by Phil Jose, Mike Gagliano, Casey Phillips, and Steve Bernocco

1. A major drawback of air purifying respirators is that:
    a. The filters must be matched to specific situations.
    b. They cannot be used in oxygen-deficient atmospheres.
    c. It is difficult to maintain a good seal.
    d. It is difficult to find the proper filters.

2. The advantage of a powered air-purifying respirator is:
    a. A blower supplies air to the user reducing the effort needed to breath.
    b. It is approved for use during structural firefighting.
    c. It maintains a better seal than a regular air- purifying respirator.
    d. It can be used in an oxygen-deficient atmosphere.

3. This type of respiratory protection is used primarily for confined space operations:
    a. air-purifying respirator
    b. powered air-purifying respirator
    c. particulate filter respirator
    d. supplied air respirator

4. A supplied air respirator (SAR) system must include for the user:
    a. an escape air cylinder
    b. a tethered communications line
    c. a buddy-breathing connection
    d. a spare particulate filter cartridge

5. Which of the following provides firefighters with an excellent level of respiratory protection while operating at structural fires?
    a. powered air-purifying respirators
    b. supplied air respirators
    c. open-circuit self-contained breathing apparatus
    d. SCUBA

6. The modern smoke environment dictates that firefighters use self-contained breathing apparatus (SCBA) whenever:
   a. Heavy amounts of smoke are present.
   b. Any level of smoke is present.
   c. A moderate amount of smoke is present.
   d. They are ordered to by an officer.

7. When burned, the smoke of this product is known to produce carcinogens, including benzene and dibenzofurans:
   a. liquefied petroleum gas
   b. hydrogen sulfide
   c. polyvinyl chloride
   d. ethyl ketone

8. This colorless and odorless smoke by-product has been increasingly cited as a cause of firefighter line-of-duty deaths.
   a. hydrogen peroxide
   b. hydrogen sulfide
   c. hydrogen chloride
   d. hydrogen cyanide

9. Firefighters should wear SCBA during overhaul activities because:
   a. Unseen toxic gases are still in the atmosphere.
   b. Carbon monoxide may be present.
   c. Particulate matter may be present.
   d. All of the above.

10. Properly worn SCBA will not protect the wearer from:
    a. chemicals absorbed through the skin
    b. facial injuries
    c. respiratory tract burn injuries
    d. inhalation of toxic chemicals

11. To ensure protection when working in a chemical, biological, radiological, or nuclear tainted environment, the firefighter should check the SCBA to be worn for:
    a. a proper fit
    b. damage to the unit
    c. A CBRN-NIOSH Agent approval label
    d. a completely full air cylinder

12. The leading cause of death for firefighters is
    a. cardiac arrest
    b. inhalation of toxic chemicals
    c. respiratory burn injuries
    d. cancer

13. For firefighters who respond to emergency scenes, medical screening should occur:
    a. every 6 months
    b. every year
    c. every 3 years
    d. only at the time of joining the department

14. High-pressure fire service air cylinders operate at
    a. 3,500 psi
    b. 4,000 psi
    c. 4,500 psi
    d. 5,000 psi

15. The most important limiting factor faced by a firefighter using an SCBA is:
    a. the pressure at which the air cylinder is rated
    b. the volume of air that is in the cylinder
    c. the type of construction of the cylinder
    d. the weight of the cylinder

16. In minutes, the three most common service times in the fire service are:
    a. 20, 40, and 60
    b. 25, 45, and 60
    c. 30, 45, and 60
    d. 30, 40, and 60

17. Minute ratings are considered to be inaccurate for the fire service because:
    a. Firefighters may consume air at a rate exceeding the 40-liter-per-minute standard.
    b. Small, allowable leaks in the system will reduce operating times.
    c. Firefighters may consume air at a rate exceeding the 35-liter-per-minute standard.
    d. Cylinder pressure ratings are considered more accurate.

18. Air cylinders should be inspected before use for damage, amount of air, and:
    a. serial number
    b. department identification sticker
    c. manufacturer's acceptance label
    d. current hydrostatic test label

19. While SCBA cylinders should be at 100% of rated capacity at all times, a cylinder may remain in service with slightly less air if:
    a. The problem is not being caused by continuing leak.
    b. The department has a written policy stating the minimum acceptable capacity.
    c. The crew working talk it over so all members know.
    d. There is a note left on the unit.

20. Cylinder construction dictates whether it needs to be _____ at a 3- or 5-year interval.
    a. hydrostatically tested
    b. visually inspected
    c. removed from service for breathing air
    d. replaced

21. Designed to transfer the load of the cylinder to the body of the user, this SCBA component may provide a handhold to remove unconscious firefighters.
    a. regulator
    b. harness
    c. back plate
    d. facepiece

22. When being used, air is provided to the firefighter through the:
    a. high-pressure hose
    b. harness
    c. regulator assembly
    d. pressure assembly

23. An existing air cylinder in an SCBA should be exchanged for a full cylinder, if the existing cylinder is at _____ of its rated capacity.
    a. 80% or less
    b. 85% or less
    c. 95% or less
    d. 90% or less

24. First-stage regulators reduce high pressure from the cylinder to approximately:
    a. 100 psi
    b. 75 psi
    c. 50 psi
    d. 30 psi

25. To be confident in the use of their SCBA, firefighters should:
    a. be able to explain all components of the SCBA
    b. know how to maintain the SCBA
    c. participate in regular and realistic training exercises
    d. wear their SCBA every time smoke is present at a fire call

26. To provide a positive pressure inside the facepiece, air is delivered at:
    a. slightly below atmospheric pressure
    b. equal to atmospheric pressure
    c. slightly above atmospheric pressure
    d. greatly above atmospheric pressure

27. Facepiece fit testing should occur:
    a. before initial training only
    b. before initial training and annually thereafter
    c. annually
    d. before initial training and every 3 years

28. The main limiting factor for a firefighter operating in an IDLH atmosphere is the:
    a. amount of air contained in the SCBA cylinder
    b. type of contaminates in the air
    c. firefighter's visibility
    d. firefighter's respiratory health

29. A trained firefighter should exit an IDLH environment:
    a. some time after the activation of the low-air warning alarm
    b. after the low-air warning alarm stops sounding
    c. immediately on the sounding of the low-air alarm
    d. before the activation of the low-air warning alarm

30. The amount of air that a firefighter uses in a given period of time is known as the:
    a. air consumption rate
    b. respiratory rate
    c. air breathing rate
    d. rate of air breathed

31. The amount of air used over a period of time by a firefighter wearing an SCBA can be affected by the firefighter's:
    a. size
    b. weight
    c. overall aerobic fitness
    d. all of the above

32. A firefighter's air consumption rate may increase in a dark, smoke-filled room because:
    a. The firefighter will slow down the pace of work being performed.
    b. More air will get used if the exit is harder to see.
    c. The firefighter has a harder time seeing the air gauge.
    d. Emotional stress increases, causing the firefighter to breath faster.

33. Fire departments need to address the _____ for firefighters operating in full personal protective equipment and SCBA
    a. completion performance times
    b. ratio of on-duty time to off-duty time
    c. ratio of work cycle to rest cycle requirements
    d. assigned functions

34. SCBA low-air warning alarms are designed to activate when the air cylinder reaches:
    a. 50% of the rated capacity
    b. 30%–35% of the rated capacity
    c. 20%–25% of the rated capacity
    d. less than 20% of the rated capacity

35. An air management plan must mandate that firefighters must exit the IDLH area before consumption of the emergency reserve begins; activation of the low-air warning is an "immediate action item for the individual and team"; and that:
    a. The low-air warning should indicate to the officer that a firefighter will be leaving.
    b. The low-air warning indicates the firefighter is using the emergency reserve.
    c. The low-air warning indicates the firefighter must complete the task and leave.
    d. The low-air warning should allow for enough air to remain to exit the building.

36. The rules of air management include know how much air you have, manage that amount of air as you go, and:
    a. Exit the IDLH area before the low-air alarm sounds.
    b. Inform your officer as to how much air you have when it gets below 50%.
    c. Be able to leave the IDLH atmosphere as soon as the alarm sounds.
    d. Know how to "buddy breathe," if required.

37. With regard to the rules of air management, a fire officer may consider assigning tasks based on:
    a. the firefighter's air consumption rate
    b. emergency breathing skills
    c. the amount of air in each firefighter's cylinder
    d. the firefighter's access to a rest and rehabilitation area

38. It is incumbent on the fire officer to:
    a. Rotate personnel based on amount of work to equalize air use by members.
    b. Identify areas of safety out of the hazard area.
    c. Determine when the team must begin to exit before the activation of low-air alarms.
    d. all of the above

39. A common cause of SCBA failure is:
    a. worn-out equipment
    b. failure to wear the unit
    c. misuse of equipment
    d. failure to provide maintenance for the unit

40. The act of buddy breathing with a civilian or other firefighter should:
    a. be practiced so members are competent if such action is needed
    b. not be condoned or practiced
    c. be included as part of an air management plan
    d. be allowed with another firefighter, but not with a civilian

41. Operating an SCBA in a low-temperature environment may cause:
    a. fogging of the mask
    b. ice to form on the cylinder
    c. high-pressure hose leaks
    d. all of the above

42. Breathing air supplied to a firefighter from an SCBA cylinder will be cooler than the surrounding air in a fire building because:
    a. Compressed air cools as it expands.
    b. Compressed air is cooled as it is pressurized during filling.
    c. The firefighter's body temperature will cool the air as it is breathed in.
    d. The SCBA cylinder does not attract and hold heat.

43. Limitations of SCBA built-in communication devices include:
    a. Transmissions may be compromised in areas with thick walls.
    b. Transmissions may be compromised if multiple units are in the same area.
    c. Feedback from units close to each other may limit the effectiveness of the units.
    d. all of the above

44. To reduce respiratory demands during a low-air emergency, all of the following are acceptable practices except:
    a. buddy breathing
    b. skip breathing
    c. breath control
    d. humming

45. An SCBA-mounted universal rescue connection is designed to be used:
    a. by rapid intervention crews for emergency filling operations only
    b. to transfer air from one firefighter's cylinder to another
    c. as a shared breathing air connection
    d. as a lifting/pulling device to help remove a downed firefighter.

CHAPTER 10 ANSWER KEY

Question #	Answer	Page #	Question #	Answer	Page #
1	b	178	31	d	191
2	a	178	32	d	191
3	d	178	33	c	191
4	a	178	34	c	191–192
5	c	179	35.	b	192
6	b	180	36	a	192
7	c	180	37	a	191–192
8	d	180	38	d	191–192
9	d	181	39	c	495
10	a	182	40	b	195
11	c	182	41	d	197
12	a	183	42	a	198
13	b	183	43	d	198
14	c	184–185	44	a	195
15	b	183	45	a	199
16	c	184			
17	a	184			
18	d	184–185			
19	b	185			
20	a	185			
21	b	185–186			
22	c	186			
23	d	185			
24	a	186			
25	c	190			
26	c	188			
27	b	188			
28	a	183			
29	d	187–188			
30	a	190			

Firefighting Tools

by Michael N. Ciampo with Rick Fritz

1. Identify the two types of hand-held, nonpowered cutting tools carried on engines and truck companies.
   a. pike head axe and hook
   b. flat head axe and bolt cutter
   c. mauls and pry bar
   d. splitting maul and sledgehammer

2. Describe the fundamental task differences between a pick-head axe and a flat-head axe.
   a. The flat head axe is heavier.
   b. The pike head axe is heavier.
   c. The flat head has a flat striking surface that can be used as a sledgehammer.
   d. The pike head axe is sharper and better for roof operations.

3. Which tools are engineered to strike other tools?
   a. spanner wrenches
   b. vice grips and pump pliers
   c. pinch point bars and pry bars
   d. mauls and flat head axes

4. Describe how a pry bar can be used to secure a ground ladder.
   a. driving the pry bar into the ground between rungs of the ground ladder
   b. placing the pry bar inside a window and securing the bar to the ground ladder with a rope
   c. placing under the heel of the ground ladder to prevent it from kicking out
   d. tying the pry bar to the inside of the ground ladder and forming a triangular brace at the base

5. Which hand tool makes a good securing post?
   a. flat head axe
   b. pry bar
   c. pike pole
   d. sledgehammer

6. Which tool is the most versatile for forcible entry tasks that require leverage?
    a. pry bar
    b. claw tool
    c. pinch bar
    d. Halligan bar

7. What is the tool of choice for making entry through a chain link fence?
    a. reciprocal saw
    b. bolt cutters
    c. rotary saw
    d. pry bar

8. Sanding, wiping, and coating with linseed oil is the process for inspecting, cleaning, and maintaining hand tools with the following type of handles:
    a. wood
    b. plastic
    c. fiberglass
    d. all types of handles

9. The definition of a single-bit axe is an axe that has only one
    a. striking head
    b. cutting blade
    c. use

10. An ideal axe head weighs
    a. 8 lb
    b. 6 lb
    c. 10 lb

11. When using an axe, on the upswing your hands should go no higher than your
    a. head
    b. waist
    c. shoulders

12. Explain the danger in swinging an axe with too much force.

13. When cutting with an axe, the blade should strike the surface
    a. straight with no angle
    b. at a slight angle
    c. so the bottom of the blade contacts only

14. The main purpose of the pick on the pick-head axe is to allow the user to
    a. make a purchase point
    b. use the axe as a prying tool
    c. pull on when removing a stuck axe.

15. The cutting surface of a bolt cutter may not be able to cut
    a. steel material
    b. aluminum material
    c. case-hardened material

16. When cutting energized electrical lines, firefighters should use
    a. bolt cutters with fiberglass handles and rubber grips
    b. dielectric bolt cutters
    c. high-quality carbon steel bolt cutters
    d. Firefighters don't cut energized electrical lines.

17. When cutting a cable or other material, why is it a good idea to know what the end result will be?

18. When cutting locks with a bolt cutter, you should always cut
    a. high on the shackle
    b. low on the shackle
    c. across the lock body

19. When disconnecting the battery power from vehicles, bolt cutters can be used to do what?
    a. Cut the battery terminals.
    b. Twist off the battery terminals.
    c. Bolt cutters should not be used.

20. The biggest difference between a flat-head axe and a splitting maul is
    a. The maul is not swung the same.
    b. The maul has two working surfaces.
    c. The maul does not cut.

21. To accommodate the off-balance head of the splitting maul, when using the tool, the firefighter should do which of the following?
    a. Shift his or her weight.
    b. Take shorter swings.
    c. Move his or her hands closer to the tool head.

22. In regard to tool location and body stance, explain the correct process when using the flat-head axe as a striking tool.

_____
_____
_____
_____
_____
_____

23. The difference between a maul and a sledgehammer is that a sledgehammer
    a. weighs less than a maul
    b. has only one striking surface
    c. has striking surfaces on both sides of the head

24. The safest way to carry a sledgehammer is by
    a. grasping the handle close to the tool head with the handle facing behind you
    b. grasping the handle close to the tool head with the handle facing in front of you
    c. grasping the bottom of the handle, allowing the tool head to be near the ground

25. Pinch-point bars and wedge-point bars are examples of
    a. striking tools
    b. prying tools
    c. cutting and striking tools

26. When used in conventional forcible entry, which tool would work more efficiently?
    a. wedge-point bar
    b. pinch-point bar
    c. Both bars would work equally well.

27. Explain how a pry bar can be used during overhaul in a plaster and lath building.

    _____
    _____
    _____
    _____
    _____

28. What is the biggest disadvantage of the claw tool?
    a. two sharp ends
    b. its overall length
    c. the lack of an engineered striking surface on the hook end

29. The best tool positioning for the firefighter using a claw tool is
    a. waist level or below
    b. waist level or higher
    c. at shoulder height

30. Explain how a claw tool can be used during overhaul to quickly remove wallboard.

_____

_____

_____

_____

_____

_____

31. For conventional forcible entry, the _____ inch Halligan bar is the preferred choice.
    a. 24
    b. 30
    c. 42

32. On the Halligan bar, the part of the tool at an immediate 90° angle to the pick is called the
    a. adze
    b. fork
    c. bevel

33. On the Halligan bar, the minimum length of the fork should be
    a. 4 in.
    b. 5 in.
    c. 6 in.

34. On the Halligan bar, the bottom side of the fork is called the
    a. dished side
    b. concave side
    c. beveled side

35. A Halligan hook is an example of a
    a. push-pull tool
    b. striking tool
    c. pulling tool

36. Explain the problems that may be encountered when using a fiberglass pike pole of 2 or more inches in diameter.

   _____
   _____
   _____
   _____
   _____
   _____

37. What tool can be more efficient than a short pike pole when working in tight areas?
    a. Halligan bar
    b. falcon hook
    c. pike axe

38. The minimum length pike pole taken into a commercial structure should be
    a. 6 ft
    b. 8 ft
    c. 10 ft

39. The most useful pike pole in commercial buildings and light industrial buildings is
    a. 8 ft long
    b. 10 ft long
    c. 12 ft long

40. This tool works well to open plaster and lath walls:
    a. New York hook
    b. Chicago pike pole
    c. Boston rake

41. When working to remove tin ceilings, the best hooks to use are the Providence hook and the
    a. New York hook
    b. Halligan hook
    c. falcon hook

42. With its angled jaw and cutting blade, this tool easily opens drywall:
    a. gypsum board hook
    b. San Francisco hook
    c. Providence hook

43. When used on the roof, the design of this tool's head allows it to push down large sections of ceiling without getting tangled in wires:
    a. rubbish hook
    b. gypsum board hook
    c. Chicago pike pole

44. A set of irons consists of which two tools?
    a. pick-head axe and Halligan bar
    b. sledgehammer and maul
    c. Halligan bar and flat-head axe

45. The tool most similar to the San Francisco bar is the
    a. Boston rake
    b. Halligan bar
    c. Chicago pike pole

46. A rubbish hook is also called an
    a. L.A. trash hook
    b. arson rake
    c. both a and b

47. This prying tool is at least 42 in. long with a fork at one end:
    a. claw tool
    b. Halligan bar
    c. pinch-point bar

48. This pike pole is used to remove plaster and lath, and it has a chisel point at the top of the pike:
    a. Boston rake
    b. San Francisco hook
    c. Providence hook

49. A fire service tool that is a true lever is a
    a. pinch-point bar
    b. Chicago patrol bar
    c. San Francisco bar

50. A pry bar with a bevel on both sides of the bar is called a
    a. beveled bar
    b. pinch-point bar
    c. wedge-point bar

CHAPTER 11 ANSWER KEY

Question #	Answer	Page #
1	b	206
2	c	207
3	d	209
4	b	210
5	b	210
6	d	211
7	b	207
8	a	216
9	b	206
10	a	206
11	c	206
12	↓	206–207

An axe swung with too much force might slip from your hands and plunge into the hole, endangering crews working below you; or you could miss your mark and hit another firefighter.

Question #	Answer	Page #
13	b	206–207
14	a	206–207
15	c	206–207
16	d	206–207
17	↓	206–207

Because cutting cables or cords may release an object being held up or in a tension situation.

Question #	Answer	Page #
18	a	207
19	b	207
20	c	208
21	a	208
22	↓	209

Hold the tool at waist level and line up the flat striking surface against the tool or object to be struck. Do not move the tool. Arrange your stance so you can effectively and strongly pivot your hips and hit the target.

Question #	Answer	Page #
23	c	209
24	b	209–210
25	b	209–210
26	a	209–210
27	↓	209–210

The firefighter can insert the pry bar into the bay of the wall, inserting the entire length of the tool, and open the wall.

Question #	Answer	Page #
28	c	210–211
29	a	211
30	↓	211

Slide the tool fork-end down into a wall opening. By grasping the hook end and pulling toward yourself, you can pull the entire length of the tool through the wall.

Question #	Answer	Page #
31	b	210–211
32	a	211
33	c	211
34	c	212
35	a	211–212
36	↓	216

They are extremely difficult to work with because they don't fit well in your hand; they are slippery when wet; and they are difficult to stow on the apparatus.

CHAPTER 11 ANSWER KEY *Continued*

Question #	Answer	Page #
37	a	207–208
38	b	212
39	b	212
40	c	213
41	c	213
42	a	213
43	a	214
44	c	216
45	b	214
46	c	214
47	a	210–211
48	b	213
49	a	210
50	c	210

Forcible Entry

by Dan Sheridan

1. Which of the following is the most accurate statement about forcible entry in the fire service?
   a. Forcible-entry techniques have not changed significantly in recent years.
   b. Forcible entry is a constantly evolving function within the fire service.
   c. The fire service is currently lagging behind modern security measures.
   d. Security measures have not changed significantly in recent years.

2. Firefighters must keep in mind that any opening that is made will also affect:
   a. fire behavior
   b. building stability
   c. overhaul operations
   d. hoseline placement

3. After forcible entry is performed, it is of the utmost importance for firefighters to:
   a. Enter immediately for extinguishment.
   b. Enter immediately for search.
   c. Remove the door from its hinges.
   d. Control the door.

4. What is the first thing firefighters should do when they enter any room?
   a. Begin a search for victims.
   b. Make sure they have a hoseline in place on the way.
   c. Check to ensure that a second way out is accessible.
   d. Give a radio report on the situation in the area.

5. When confronted with wooden doors that have panels that are easily removed, under what conditions would it be acceptable to remove the panels instead of forcing the door?
   a. when there are additional heavy security devices on the door
   b. when the fire condition is heavy and water can be applied through the opened panel to knock down the fire
   c. when the emergency is minor
   d. when a known exit is located on the opposite side of the apartment

6. This forcible-entry tool has a chisel on one end that can drive outer rim locks once the cylinder is removed:
   a. A-tool
   b. rex tool
   c. K-tool
   d. duckbill

7. This forcible-entry tool has only one function, to go through the lock:
   a. A-tool
   b. rex tool
   c. K-tool
   d. duckbill

8. This forcible-entry tool combines well with a maul or the back of an axe:
   a. A-tool
   b. rex tool
   c. K-tool
   d. duckbill

9. The rabbit tool is typically sued to force which type of doors?
   a. inward-swinging metal doors
   b. outward-swinging metal doors
   c. overhead rolling metal doors
   d. iron window gates

10. This type of steel should not be cut with bolt cutters:
    a. hot-molded
    b. cast
    c. rolled
    d. case-hardened

11. This is the secret to forcible entry:
    a. experience
    b. proper tools
    c. self-confidence
    d. paying attention to the obvious

12. What is the key to getting past a lock?
    a. understanding vulnerabilities and attacking at those points
    b. applying force on the proper position on the tool
    c. applying force on the proper position on the lock
    d. first examining the lock from a distance

13. When forcing a heavy-duty padlock, which of the following is the preferred tool?
    a. duckbill
    b. power saw with metal blade
    c. heavy-duty bolt cutters
    d. rabbit tool

14. American Lock™ Series 2000 locks should be forced using which tool(s)?
    a. duckbill
    b. power saw with metal-cutting blade
    c. bolt cutters
    d. maul and Halligan tool

15. This type of door lock is cut into and thus part of the door frame:
    a. rim lock
    b. cylindrical lock
    c. mortise lock
    d. fox lock

16. This type of door lock is surface-mounted on the inside of the door:
    a. rim lock
    b. cylindrical lock
    c. mortise lock
    d. fox lock

17. This type of door-locking mechanism is contained in the door's lever or knob:
    a. rim lock
    b. cylindrical lock
    c. mortise lock
    d. fox lock

18. This type of locking device has two bars that hold the door closed from the inside:
    a. rim lock
    b. cylindrical lock
    c. mortise lock
    d. fox lock

19. In order to force a fox lock, what must you do first?
    a. Remove the vertical bolt and the mating plate.
    b. Shear three of the four bolts and expose the cylinder.
    c. Use a 5/32 square tool to open the locking mechanism.
    d. Turn the cylinder toward the bottom two bolts while pushing toward the door.

20. All magnetic locks work only with a(n) _____ power supply
    a. alternating current (AC)
    b. direct current (DC)
    c. generator current (GC)
    d. indirect current (IC)

21. Which of the following is the most common type of electric strike?
    a. fail-secure
    b. electromagnetic
    c. fail-safe
    d. fail-proof

22. This type electric strike stays locked even without power:
    a. fail-secure
    b. electromagnetic
    c. fail-safe
    d. fail-proof

23. Outward-swinging doors with magnetic locks should be forced using which tool(s)?
    a. screwdriver
    b. hydraulic spreading devices
    c. K-tool
    d. flat-head axe and Halligan tool

24. Inward-swinging doors with magnetic locks should be forced using which tool(s)?
    a. screwdriver
    b. hydraulic spreading devices
    c. K-tool
    d. flat-head axe and Halligan tool

25. The majority of the doors that we encounter are this type:
    a. sliding
    b. rolling
    c. inward-swinging
    d. outward-swinging

26. This type of door glass, if smashed, will explode into little pieces:
    a. plate
    b. tempered
    c. wired
    d. Lexan®

27. Where is the locking mechanism usually found in tempered glass doors?
    a. in the top stile
    b. in the bottom stile
    c. in the stile on the hinge side
    d. in the stile on the handle side

28. Regarding forcible entry at fires, why do we need to attack the normal point of entry?
    a. It is closest to the arrival position of the apparatus.
    b. The door is not usually locked.
    c. It is the route most people take in a fire.
    d. The light is usually best compared to other areas of the structure.

29. When holding a Halligan tool, your hands should never be anywhere but:
    a. on the shaft
    b. on the adze and shaft
    c. on the fork and the shaft
    d. on the adze and the fork

30. First and foremost, what action should be taken when forcing a door?
    a. Feel the door for heat.
    b. Size up the door to determine the best means possible of entry.
    c. Vent nearby windows to release heat.
    d. Set up lights in the area.

31. To control an inward-opening door, what should firefighters do prior to forcing the door?
    a. Use a rope or hose strap to hold the door.
    b. Remove the hinge pins to disable the swing of the door.
    c. Place a wooden wedge beneath the door to keep it from moving.
    d. Ensure that they can enter as soon as the door is forced.

32. When using the hydraulic spreading tool on an inward-opening door, where is best location to place the jaws of the tool?
    a. 1 ft below the lock
    b. 1 ft above the lock
    c. directly on the lock
    d. at the top of the door

33. Through-the-lock techniques are normally used with these locks (identify two):

    1. flush lock
    2. rim lock
    3. recessed lock
    4. mortise lock

    a. 1 & 3
    b. 2 & 3
    c. 1 & 4
    d. 2 & 4

34. You are removing a mortise lock. You have removed the cylinder and found that the bolt is at the 5 o'clock position. What do you do to spring the lock?
    a. Move the bolt to the 6 o'clock position.
    b. Move the bolt to the 12 o'clock position.
    c. Move the bolt to the 7 o'clock position.
    d. Pull the bolt out of the door with a Vise-Grip® tool.

35. You have pulled the cylinder of a rim lock and found that a metal shutter has blocked your key tool from manipulating the locking mechanism. What do you do?
   a. Use the point of the Halligan tool and knock the lock off the door.
   b. Use a Vise-Grip tool to pull the shutter back.
   c. Slide the shutter to the 12 o'clock position.
   d. Use the hydraulic spreader to force the door.

36. When forcing an inward-opening door with a flat-head axe and Halligan tool, where do you position the Halligan tool?
   a. directly on the lock
   b. 6 in. above or below the lock
   c. behind the lower hinge pin
   d. behind the upper hinge pin

37. What is the best method of forcing a residential overhead garage door?
   a. Use a power saw and cut around the lock.
   b. Break out a panel near the lock.
   c. Pry up the bottom of the door with a Halligan tool or pry bar.
   d. Break out a panel near the corner.

38. When confronted with roll-down gates, the primary point of attack and usually the fastest way to gain entry is to:
   a. Cut the padlocks.
   b. Use a slash cut.
   c. Pry the door out of its tracks.
   d. Use an inverted V-cut.

39. Regarding roll-down gates, this type of cut allows for quick stream penetration at ceiling level prior to entry for a quick knockdown:
   a. box cut
   b. slash cut
   c. inverted V-cut
   d. cone cut

40. Regarding plate glass windows, firefighters have often underestimated this, and it has led to injury:
   a. their resistance to breaking
   b. their weight
   c. the toxicity of the shards
   d. their size

41. This type window has an impact resistance 250 times greater than safety glass:
    a. wired glass
    b. Lexan
    c. casement
    d. tempered

42. You are making entry through a double-hung window into an apartment to search. For safety reasons, what is the best action?
    a. Place a tool in the window to prop it open.
    b. Ensure that a company is coming from the opposite way with a hoseline.
    c. Close the window so as not to draw fire to your position.
    d. Remove both the bottom and top sashes.

43. The tools of first choice for burglar bar removal are (choose two):
    1. sledgehammer
    2. Halligan tool and flat-head axe
    3. air chisel
    4. rotary saw with metal-cutting blade

    a. 1 & 3
    b. 2 & 3
    c. 3 & 4
    d. 1 & 4

44. When using a rotary saw on burglar bars, this should be a last resort:
    a. Cut through the main bars.
    b. Remove the moving sections of the hinges.
    c. Cut through the dead bolt.
    d. Cut where the bars are flattened and attached to the building.

45. These are the "master keys" for all-purpose forcible entry on the fireground:
    a. K-tool and A-tool
    b. forcible-entry irons
    c. air chisel and Sawzall®
    d. power saw with metal-cutting and carbide-tip blades

46. When confronted with window gates, after opening the window, what would be the next action to take?
    a. Kick the gate in.
    b. Force the top hinge.
    c. Force the bottom hinge.
    d. Slide the gate toward the lock.

47. When searching a large fire area, it is a good idea to do this:
    a. Vent all the windows available.
    b. Leave your partner at your entry point.
    c. Establish a second means of egress.
    d. Enter only after water has started being applied to the fire.

48. You are looking to force hurricane-resistant windows. Your saw is out of service. What tool would you use as your next choice?
    a. air chisel
    b. hydraulic spreading tool
    c. pruning shears
    d. sledgehammer

49. Suppose you are conducting a wall breach operation and come across a double-header beam. What action should you take?
    a. Do not remove the beam because a collapse could occur.
    b. Remove the beam. It is not dependent on the materials below.
    c. Only remove one of the beams because removing both could be dangerous.
    d. Check the ceiling above to determine how far the beam runs.

50. You are breaching a masonry wall with a battering ram. What shape would you create in making the hole?
    a. V shape
    b. O shape
    c. T shape
    d. upside-down V shape

CHAPTER 12 ANSWER KEY

Question #	Answer	Page #	Question #	Answer	Page #
1	b	234	31	a	248
2	a	234	32	c	248
3	d	234	33	d	249
4	c	234	34	c	249
5	c	234	35	a	249
6	b	235	36	b	248–249
7	c	235	37	b	250
8	d	235	38	a	251
9	a	235–236	39	d	251
10	d	236	40	b	253
11	c	238	41	b	253
12	a	238	42	d	254
13	b	239	43	c	256
14	b	240	44	a	257
15	c	240	45	b	257
16	a	241	46	b	258
17	b	241	47	c	248
18	d	241	48	c	259
19	b	242	49	a	259
20	b	243	50	d	260–261
21	a	243			
22	a	243			
23	d	243			
24	b	243			
25	c	243			
26	b	244			
27	b	244			
28	c	247			
29	a	247			
30	b	247			

Ladders

by Michael Ciampo

1. An apparatus having a permanently mounted telescoping ladder operated by a hydraulic fluid and lift system, in conjunction with steel cables and pulleys, best describes:

    a. an articulating ladder
    b. a tower ladder
    c. an aerial ladder
    d. a ladder tower

2. An apparatus equipped with telescoping boom sections attached to a bucket, a pre-piped waterway, and a monitor nozzle, is best known as:

    a. a quint apparatus
    b. an articulating ladder
    c. a tower ladder
    d. an aerial ladder

3. A quint fire apparatus is equipped with a permanently mounted fire pump, an aerial ladder with a waterway, a water tank, portable ground ladders and:

    a. a hose storage compartment
    b. a mounted generator and lighting
    c. a foam tank
    d. a ladder platform or bucket

4. A single section wall ladder is also known as:

    a. an attic ladder
    b. a straight ladder
    c. a Fresno ladder
    d. a folding ladder

5. A straight ladder with permanently attached, spring-loaded curved metal hooks is better known as:

    a. a wall ladder
    b. a scissor ladder
    c. a combination ladder
    d. a roof ladder

6. A ground ladder with a bed section and one or two fly sections is best termed:
   a. a combination ladder
   b. an A-frame ladder
   c. a scaling ladder
   d. an extension ladder

7. A narrow extension ladder that is commonly called a two-section attic ladder, best describes:
   a. a Fresno ladder
   b. a Pompier ladder
   c. an A-frame ladder
   d. a closet ladder

8. A ladder constructed of one center beam with rungs attached to each side and a large forged hook at the top of the beam, best describes:
   a. a roof ladder
   b. a Fresno ladder
   c. a Pompier ladder
   d. a combination ladder

9. This ladder must be extended manually, and its components include two pins and two receiver brackets mounted at the ladder tip.
   a. a Fresno ladder
   b. an A-frame ladder
   c. an attic ladder
   d. a roof ladder

10. Wood and fiberglass ladders are considered nonconductive ladders because they:
    a. have few maintenance requirements
    b. are constructed in a similar manner
    c. do not attract electricity even when wet
    d. do not attract electricity when dry

11. The main structural component of a ladder that supports a firefighter's weight and transfers it to the ground is called the:
    a. beam
    b. rung
    c. butt
    d. bed

12. The butt of the ladder can also be referred to as the:
    a. heel
    b. foot
    c. base
    d. both a & c are correct

13. The dog or pawl on a ladder is a:
    a. slot that supports and interlocks with a corresponding section of a ladder as it is raised
    b. spring-loaded locking device
    c. slip-resistant pivoting safety shoe
    d. spurs or spikes mounted onto a ladder at its base

14. Electrical hazard warning labels can usually be found affixed to a portable ladder:
    a. at the bottom of the beam near the base
    b. at the top of the beam near the tip
    c. on the base section between the fourth and fifth rungs
    d. on the fly section between the second and third rungs

15. The fly section of a ladder is best described as:
    a. the section of the ladder that remains in contact with the ground
    b. the section of the ladder that extends to gain height and distance
    c. the main component that transfers the climber's weight to the ground
    d. the section on the ladder where it evenly balances the ladder when it is lifted

16. Heat sensor labels turn black to warn users that the ladder has been exposed to enough heat to damage it. At what temperature does this occur?
    a. above 200 degrees Fahrenheit
    b. above 300 degrees Fahrenheit
    c. above 400 degrees Fahrenheit
    d. above 500 degrees Fahrenheit

17. The rope used to raise the ladder's fly section out of the bed section is known as the:
    a. tie-off line
    b. securing rope
    c. lashing line
    d. halyard

18. Normally found only on wooden ladders, this ladder component runs through both beams to help secure the beams and rungs:
    a. tie rods
    b. rung rods
    c. beam rods
    d. ladder rods

19. Tormentor poles on portable ladders are used for:
    a. helping secure the ladder on the apparatus
    b. helping secure the ladder to the structure
    c. helping firefighters stabilize the ladder as it is raised and lowered
    d. helping firefighters get on to and off of the raised ladder at the roof level

20. Portable ladders should be inspected after each use prior to be being placed back on the apparatus _____:
    a. every week
    b. every month
    c. every other month
    d. every three months

21. While conducting an inspection of a roof ladder, you should check that roof hook tension springs operate correctly, the hook's shape is not distorted, the hooks operate freely, and:
    a. the ladder locks function properly
    b. the pulley operates freely
    c. the hook ends are sharp
    d. the halyard is not frayed

22. The serviceability of a portable ladder should be checked annually, using which of the following methods:
    a. deflection recovery test
    b. climbing angle load test
    c. vertical weighting test
    d. horizontal load test

23. After cleaning a wooden ladder, it is important to find and repair any areas of damaged varnish because the varnish:
    a. helps keep the paint below it in good condition
    b. helps maintain the ladder's nonconductivity
    c. makes it quicker to remove debris when cleaning the ladder
    d. makes it easier to slide the beams through your gloved hands when raising or lowering the ladder

24. To easily identify the lengths of ladders when they are stored on the apparatus, firefighters should:
    a. label the base of the ladder butts with the ladder size
    b. label the beams at the top and bottom with the ladder size
    c. attach a sign adjacent to the ladder storage area listing ladder sizes
    d. color code ladder brackets

25. "Marking the rung two up from the center rung with a small paint mark or piece of electrical tape near the beams or in the center of the rung," describes how a firefighter can best identify a ladder's:
    a. hoisting point
    b. balance point
    c. high point tie-off
    d. flex point

26. When operating with portable ladders where electrical power lines are present, firefighters should maintain a minimum distance of _____ feet from the lines.
    a. four
    b. six
    c. eight
    d. ten

27. When placing ladders on a structure, the ladders should not be placed in front of lower floor windows or doorways because:
    a. Fire may come out of the window, cutting off the escape route for the firefighters using the ladder.
    b. A firefighter exiting the building may run into the ladder, causing it to fall.
    c. A hose line, when charged, may move and knock over the ladder.
    d. all of the above

28. Prior to choosing a ladder to use when at a fire scene, it is important to determine where the ladder must be placed, what length of ladder is needed, and:
    a. whether the ladder is damaged
    b. what the ladder is constructed of
    c. what the purpose of the ladder is
    d. what the construction of the building is

29. Commercial structures usually measure _____ feet from floor to floor:
    a. 8–10
    b. 10–12
    c. 12–14
    d. 15–16

30. When removing a roof ladder from a pumper to reach an extension ladder, the best location to place the roof ladder is:
    a. back on the ladder brackets
    b. under the apparatus
    c. next to the apparatus, flat on the ground
    d. on the side of the apparatus leaning up against the rear tire

31. When removing a portable ladder from the rear compartment of an aerial apparatus, the firefighter should be aligned "body to the building." This means the firefighter should be:
    a. positioned to maintain visual contact with the apparatus driver
    b. positioned on the side of the portable ladder that allows for the fire building to be in sight
    c. positioned on the side of the portable ladder with his or her back to the fire building
    d. positioned so the portable ladder will always pull out straight towards the fire building

32. When moving a roof ladder up an aerial ladder on a steep incline, the best option for the firefighter is to:
    a. Place the roof ladder on its beam and slide it up the rungs of the aerial ladder.
    b. Lay the roof ladder across both rails of the aerial ladder and slide it up the rails.
    c. Secure the roof ladder in a shoulder-carry position and climb the aerial ladder using a hand-over-hand motion on the rungs.
    d. Open the roof ladder hooks and push the ladder up each rung of the aerial ladder, while using the hooks to secure the roof ladder to each rung of the aerial while climbing.

33. To assist firefighters climb down over parapet walls to reach the roof of a structure, some aerial ladders may be equipped with a(n) _____ inside the tip of the aerial.
    a. extension ladder
    b. roof ladder
    c. folding ladder
    d. A-frame ladder

34. When raising a ladder on uneven terrain, which of the following could be used to help stabilize the portable ladder?
    a. chocks or cribbing
    b. trenching
    c. blocks or bricks
    d. all of the above

35. To achieve the proper climbing angle for a portable ladder, the base of the ladder should be placed:
    a. one-half the total working length of the ladder away from the building
    b. one-third the total working length of the ladder away from the building
    c. one-fourth the total working length of the ladder away from the building
    d. one-fifth the total working length of the ladder away from the building

36. When raising an extension ladder, the firefighter assigned to the butt end should:
    a. operate the halyard to raise the ladder while positioned between the ladder and the building
    b. operate the halyard to raise the ladder while positioned in front of the ladder
    c. help balance the ladder while the second firefighter operates the halyard
    d. watch the extending fly section while helping to balance the ladder

37. The preferred method for footing a ladder should have a firefighter positioned:
    a. behind the ladder, with his or her back to the building
    b. facing the ladder, in the direction of the climb or building
    c. holding onto the right beam while slightly in front of the ladder
    d. holding onto the left beam while slightly behind the ladder

38. If the climbing firefighter is operating off of one side of the ladder in a leg-lock position, the footing firefighter should:
    a. Move away from the ladder so as not be struck by falling glass or debris.
    b. Move to the same side of the ladder as the climbing firefighter, while pushing against the ladder beam.
    c. Maintain their current position, so as not to lose a grip on the ladder.
    d. Place his or her foot on the opposite side of ladder's base.

39. Moving an extended portable ladder after it has been positioned on a building is best accomplished by:
    a. retracting the fly section, lowering the ladder, and re-raising at the desired location
    b. sliding it across the face of the structure
    c. rolling it across the face of the structure
    d. with the aid of a second firefighter, bring the ladder vertical and carry it to the new location

40. When climbing an aerial ladder, firefighters should space themselves _____ feet apart:
    a. 6
    b. 8
    c. 10
    d. 12

41. When climbing a portable ladder with a saw clipped to a carrying sling, the firefighter should use which of the following methods:
    a. The sling should be placed over the head, with the saw resting underneath the arm and back.
    b. The sling and saw should be pushed farther back and on top of the SCBA cylinder.
    c. The sling and saw should be placed over a shoulder.
    d. Any of the above methods may be used.

42. When two firefighters are assigned to advance an uncharged hoseline up a ladder, the job of the bottom firefighter is to:
    a. Remain on the ground, foot the ladder, and feed the hose to the climbing firefighter.
    b. Remain on the ground and foot the ladder.
    c. Remain on ground and feed hose to the climbing firefighter.
    d. Assist in advancing the hoseline by climbing up the ladder behind the first firefighter.

43. When working from an aerial ladder, a firefighter should secure himself or herself by:
    a. tying onto the ladder with the use of a utility rope
    b. wearing and using an approved safety belt
    c. using a leg-lock
    d. having a second firefighter maintain contact on the working firefighter

44. A firefighter is assigned to lower the drop ladder for a fire escape on the outside of a building. Before reaching up with a tool to release the ladder, the firefighter should be positioned:
    a. directly under the drop ladder
    b. on the same side of the fire escape as the ladder, and in front of the drop ladder
    c. under the landing
    d. in front of the landing and immediately adjacent to the side of the drop ladder

45. "Reach the proper working height on the ladder with both feet on the same rung; place a boot in the center of the rung with the heel slid into the rung; next, lift the other boot over the rung and through the rung spacing." These instructions best describe the:
    a. use of a ladder safety belt
    b. leg-lock maneuver
    c. hyperextend leg-lock maneuver
    d. hook-in leg-lock maneuver

46. When using the arm-lock maneuver to vent a window off of the right side of the ladder, the firefighter's left hand should be placed:
    a. on the left beam of the ladder
    b. between the two rungs directly in front of the firefighter and gripping the tool butt
    c. on the rung directly in front of the firefighter
    d. on the tool being used, directly next to the right hand

47. The first step when faced with having to remove a conscious victim from a building and down a portable ladder is to:
    a. Make verbal contact with the victim.
    b. Raise a ladder of proper height to the windowsill.
    c. Raise a ladder of proper height to the side of the window.
    d. Check the ground condition for ladder placement.

48. When removing a semiconscious victim down a ladder, the firefighter should place one knee into the buttocks of the victim for support and place his or her hands:
    a. always at one side of the victim and on one rung at a time while descending
    b. so the victim's wrists are securely between the firefighter's hands and the ladder beams
    c. around the victim and onto the back of the ladder beams
    d. around the victim and onto the ladder rungs

49. When removing an unconscious victim down a portable ladder, the firefighter can control the descent by:
    a. placing the victim onto his or her shoulder
    b. by using his knees to balance the victim's weight
    c. pressing his legs into the inside beams of the ladder
    d. pressing the victim into the ladder with his or her chest

50. In order to check the sturdiness of the structure supporting the tip of the ladder, the firefighter should:
    a. drop the ladder against the fire building on the area in question
    b. gently press on the ladder at about chest level
    c. partially climb the ladder, hold both beams, lean back slightly and "bounce" the ladder against the building
    d. partially climb the ladder and use a pike pole to tap the supporting wall on both sides of the ladder tip

51. Portable ladders for accessing roofs should have at least five rungs above the roofline except when:
    a. situations occur where the ladder may interfere with operations, such as short porch roofs with varying grades of slope
    b. roof ladders need to be deployed
    c. parapet walls enclose the edge of the roof
    d. light beacons are attached to the ladder

52. After completely removing a window for entry into a building from a portable ladder, the next step the firefighter must take is to:
    a. check the floor conditions with a sweep and sound maneuver
    b. check the floor conditions by sounding the floor with a tool
    c. check the floor conditions by entering feet first and probing with a foot
    d. check the floor conditions by dropping a tool onto the floor

53. When placing a portable ladder to remove people from an exterior balcony, the best position for the ladder is:
    a. against the balcony, with the tip placed at the same height of the railing
    b. against the balcony, with the tip placed several rungs above the railing
    c. against the building, with the ladder beam adjacent to the railing and the tip several rungs above the railing
    d. against the building, with the ladder beam adjacent to the railing and the tip at the same height of the railing

54. When conducting ladder drags, the main difference between the Swick and Creed methods is:
    a. the use of a longer extension ladder
    b. the positioning of the roof ladder
    c. the hand tools used
    d. how the halyard is tied

55. When using an extension ladder to vent a window, just prior to the ladder striking the glass, the firefighter should:
    a. Warn others in the area of possible falling glass.
    b. Get on the bottom rung so his or her body weight will assist in breaking the window.
    c. Make sure that the glass will be hit at the center point.
    d. Remove his or her hands from the ladder to avoid broken glass sliding down the beams.

56. Firefighters raising a three-section extension ladder need to be concerned with:
    a. ladder weight and bulkiness
    b. getting both fly sections to lock into place
    c. the ladder is raised by 28 inches rather than 14 inches when the halyard is pulled down
    d. All of the above are real concerns.

57. A large extension ladder that has permanently attached support poles hinged and mounted on each side of the bed section is called a:
    a. Bangor ladder
    b. tormentor ladder
    c. pole ladder
    d. both a & c are correct

58. Support poles on Bangor ladders are used to do all of the following except:
    a. raising the ladder
    b. carry the ladder's weight
    c. lowering the ladder
    d. stabilizing the ladder

59. Which of the following ladders could be used in a stretcher-like fashion to help remove an injured firefighter through a basement window:
    a. attic ladder
    b. A-frame ladder
    c. roof ladder
    d. stepladder

60. If a sheet metal ridge vent prohibits the hooks of a roof ladder from being properly secured, a firefighter may need to:
    a. Support the butt end of the roof ladder against the raised fly section of the ground ladder at the roofline.
    b. Work on the roof without the roof ladder in place.
    c. Use a Halligan bar to make two holes in the opposite side of the ridge for the roof ladder hooks to sit in.
    d. Tie the roof ladder onto the chimney or other roof mounted object.

CHAPTER 13 ANSWER KEY

Question #	Answer	Page #	Question #	Answer	Page #
1	c	266	31	b	280
2	c	266	32	c	281–282
3	a	267	33	c	282
4	b	267	34	d	287
5	d	267	35	c	288
6	d	267	36	a	288
7	a	268	37	b	286
8	c	268–269	38	d	286
9	b	268–269	39	c	287
10	d	269	40	c	289
11	a	269	41	d	290
12	d	271	42	a	290–291
13	b	271	43	b	291
14	c	272	44	c	293
15	b	272	45	d	294
16	b	272	46	b	296
17	d	272	47	a	297
18	a	273	48	c	298
19	c	273	49	d	298
20	b	273	50	b	299
21	c	273–274	51	a	299
22	d	274	52	a	300
23	b	274	53	c	301
24	a	275	54	b	303
25	a	276	55	d	314
26	d	276	56	d	315
27	d	276	57	d	316
28	c	277	58	b	316
29	b	277	59	a	317
30	a	277	60	c	311

Ventilation

by John Mittendorf with Michael Ciampo
and Kurt Zingheim

1. In most structure fires, ventilation:
   a. should always be performed
   b. should be done in the upper most floors first
   c. should never be performed
   d. should be a primary concern

2. Fire gases generally:
   a. spread out horizontally
   b. spread vertically
   c. spread in all directions equally
   d. rise vertically to the ceiling and then spread horizontally across the underside of the ceiling

3. Which of the following statements is correct?
   a. Only large fires produce smoke and fire gases that require ventilation.
   b. Smoke from a smoldering fire does not need to be ventilated.
   c. Horizontal, vertical, or pressurized ventilation can improve visibility and reduce dangerous concentrations of heat, smoke, and fire gases.
   d. Entering an unvented fire poses little risk to firefighters.

4. Smoke is:
   a. a mixture of carbon monoxide and hydrogen cyanide
   b. composed of water vapor and unburned particulate matter
   c. the by-product of complete combustion
   d. composed of solid and liquid particulates (aerosols) and fire gases

5. The most common fire gas with the highest toxicity is:
   a. carbon monoxide
   b. hydrogen cyanide
   c. carbon dioxide
   d. hydrogen chloride

6. A bulkhead is:
   a. an extension of a stairwell shaft above the roof
   b. a hatch in the roof of a building
   c. the portion of a fire wall rising above the roofline
   d. another name for a parapet wall

7. A scuttle is:
   a. an extension of a stairwell shaft above the roof
   b. a small, covered opening providing access to the roof
   c. the portion of a fire wall rising above the roofline
   d. another name for a parapet wall

8. When fighting a first floor fire in a five story building, you should:
   a. Vent by cutting open the roof.
   b. Vent by cutting open the roof after breaking windows.
   c. Only open windows on the fire floor.
   d. Open the bulkhead or scuttle to relieve the stairwell of smoke and fire gases.

9. What type of roof is normally spongy?
   a. clay tile
   b. inverted built-up roof
   c. wood shingle
   d. asphalt shingle roof

10. The two types of mechanical ventilation are normally classified as:
    a. hydraulic ventilation and forced air ventilation
    b. fan-assisted ventilation and HVAC ventilation
    c. horizontal and vertical ventilation
    d. negative pressure ventilation and positive pressure ventilation

11. When a contaminant fills a room:
    a. The lighter, cooler gases rise to the top, and the heavier, hotter gases settle to the floor.
    b. The lighter, hotter gases rise the top, and the cooler, heavier gases settle to the bottom.
    c. Fire gases are diffused throughout the room, creating thermal balance.
    d. The smoke rises to the top, but the heat radiates out equally.

12. Negative pressure ventilation involves placing a fan or blower:
    a. outside a window or doorway to pull smoke and gases out
    b. inside a window or doorway to exhaust smoke and gases to the exterior
    c. inside a room to push fresh air into the room
    d. outside a door to push fresh air into the room

13. Positive pressure ventilation involves placing a fan or blower:
    a. outside a window or doorway to pull smoke and gases out
    b. inside a room to exhaust smoke and gases to the exterior
    c. inside a room to push fresh air into the room
    d. outside a door to push fresh air into the room

14. The top priority in any ventilation operation is:
    a. improving visibility for firefighters
    b. safety
    c. improving the interior environment to aid in evacuation and increase a victim's chance of survival
    d. limiting damage to the structure

15. Removing screens during horizontal ventilation
    a. is not necessary if a fan or blower is used
    b. can be done but is not vital
    c. should be done because screens restrict air movement by 25%
    d. is vital because screens can restrict air movement by at least 50%

16. All of the following building features can be used for ventilation purposes except:
    a. vent pipes
    b. scuttles
    c. bulkheads
    d. dumbwaiter shafts

17. Trimming a window when venting means:
    a. installing decorative trim
    b. removing decorative trim t expose window sashes
    c. removing glass panes without breaking them, by peeling away weather stripping
    d. using a hand tool to go around the outside of a window to remove leftover glass shards of glass from the window

18. An extension type ground ladder can be used to vent an upper story window.
    a. False; ground ladders are never used as a vent tool.
    b. True; but only when lying flat.
    c. True; but only from the beam.
    d. True; it may be done from the flat or beam position.

19. The arm-lock maneuver for venting windows is accomplished by:
    a. placing both arms outside the ladder's beams and grasping the tool
    b. placing both arms inside the beams and through the rungs, then grasping the tool
    c. locking one arm around a rung and swinging the tool with the free hand
    d. placing one arm through the rungs, grasping the butt of the tool, and holding the tool outside the beam with the free hand

20. When venting hurricane-resistant glass windows the tool(s) of choice is/are:
    a. Halligan bar and a flat-head axe
    b. reciprocal saw
    c. chainsaw with a carbide-tip blade
    d. hydraulic ram tool

21. The presence of 2 × 4 cross members over a piece of plywood or oriented strand board (OSB) indicates of what method of window covering?
    a. Philadelphia method
    b. HUD method
    c. Atlantic City method
    d. Urban Institute method

22. When opening a window covered with plywood or OSB and secured in place with 2 × 4 lumber and bolts, you should plunge a chainsaw through the board deep enough to cut the interior securing 2 × 4.
    a. True; but you need to cut the bottom first.
    b. True; but you need to cut the top first.
    c. True; but you should do so at a 45° angle.
    d. False; doing so could cause the saw to hit glass, a window sash, or other obstacles and potentially injure any firefighters or victims in close proximity to the window.

23. When performing a vent, entry, and search (VES) technique, a firefighter should probe the floor with a tool first, then:
    a. Dive in headfirst to enter as fast as possible.
    b. Slide in feetfirst to make sure the floor is solid.
    c. Enter with one leg, then the other, and remove all curtains, blinds, and obstructions.
    d. Remove all curtains, blinds, and obstructions, and enter with one leg, then the other.

24. All of the following are reasons for closing the room door when doing the VES technique except:
    a. It will inform interior search teams that the room has already been searched.
    b. It decreases the chance of pulling the fire into the room.
    c. It may increase visibility, making it easier to find a victim.
    d. It creates a safety factor, with the door holding back the fire while searching the room.

25. The preferred order of accessing a flat roof on a multiple dwelling is:
    a. fire escape, tower ladder, ground ladder
    b. alternate interior stair if available, adjoining building, fire escape, tower ladder
    c. alternate interior stair if available, adjoining building, tower ladder, fire escape
    d. tower ladder, fire escape, adjoining building

CHAPTER 14 ANSWER KEY

Question #	Answer	Page #
1	d	324
2	d	324
3	c	324
4	d	325
5	b	326
6	a	327
7	b	327
8	d	327
9	b	328
10	d	330
11	b	330
12	c	330
13	d	331
14	b	371
15	d	334
16	a	340
17	d	341
18	d	341–342
19	d	342
20	c	344
21	b	346–347
22	d	347–348
23	d	348
24	a	350
25	c	356–357

Water Supply and Hose

by Dennis LeGear

15

1. One of the first critical elements of all successful fire attacks is _____.
   a. locating the fire in a timely manner
   b. ensuring the cause of the fire has been properly determined
   c. ensuring that radio communications are disciplined
   d. establishing an adequate water supply

2. The weight of the air that is exerted on objects on the earth is known as _____.
   a. gravity
   b. circumferential deference
   c. atmospheric pressure
   d. barometric pressure

3. The loss of pressure and energy that occurs when water is flowing is termed _____.
   a. inverse pressure
   b. friction loss
   c. reduced flow velocity
   d. Boyle's law

4. If you took a pressure reading at a closed nozzle with no water flowing, what type of pressure would you be measuring?
   a. static pressure
   b. linear pressure
   c. residual pressure
   d. flow pressure

5. This measurement represents the pressure remaining in the water supply system when water is flowing:
   a. static pressure
   b. linear pressure
   c. residual pressure
   d. flow pressure

6. When a hoseline is closed and no water is flowing, what type of energy exists in the water in the hose?
    a. kinetic energy
    b. linear energy
    c. potential energy
    d. flow capability

7. Violent dissipation of kinetic energy that occurs when flowing water is stopped rapidly in a closed system is termed _____.
    a. inverse flow
    b. friction loss
    c. water hammer
    d. gravity disruption

8. The best practice in preventing water hammer is by:
    a. keeping pumping equipment in proper operating condition at all times.
    b. keeping hose stretches short and couplings tight
    c. operating valves and nozzles in a slow, deliberate, and cautious manner
    d. ensuring that pump pressures do not exceed manufacturer's recommendations

9. Aquifers are an example of what type of water source?
    a. groundwater source
    b. static water source
    c. surface water source
    d. natural water source

10. A river is an example of what type of water source?
    a. groundwater source
    b. static water source
    c. surface water source
    d. natural water source

11. In a water system, the amount of water treatment that is required is directly related to _____.
    a. the amount of water available
    b. the static pressure of the surface water in the area
    c. how clean the raw water entering the system is
    d. when the water will be needed for consumption

12. The most important part of a municipal water system in regard to fire departments is the _____.

    a. treatment and storage facilitates
    b. distribution system
    c. hydrants
    d. need to pump the water long distances

13. Which is not one of the ways in which water distribution systems are designed to provide low friction loss and good residual pressure during normal use?

    a. Water main valves are located throughout the system at frequent intervals, so repairs and maintenance can be conducted without affecting the whole system.
    b. A series of small-diameter pipes are enlarged to medium-diameter pipes and then enlarged even further to large-diameter pipes that are laid out in a gridlike fashion.
    c. Water is received at most locations from multiple directions from different sized mains.
    d. A series of large-diameter pipes are reduced to medium-diameter pipes and then reduced even further to smaller-diameter pipes that are laid out in a gridlike fashion.

14. Which of the following is not true regarding water main valves?

    a. Dead-end water main valves are usually kept in the closed position.
    b. Most water main valves are always left in the open position.
    c. Most water main valves are commonly spaced at 500 and 800 feet.
    d. Water main valves can be kept in a closed position for the purpose of separating grids or pressure zone.

15. These make up the bulk of most water main systems and generally range in size from the outdated 4-in. main to 12-in. mains. Hydrants are primarily installed on this portion of the system.

    a. primary feeders
    b. secondary feeders
    c. tertiary feeders
    d. distributors

16. These mains are typically 30–48 in. in diameter and have no hydrants connected to them:

    a. primary feeders
    b. secondary feeders
    c. tertiary feeders
    d. distributors

17. These water mains are generally 12–24 in. in diameter and are common throughout most municipal water main systems:
    a. primary feeders
    b. secondary feeders
    c. tertiary feeders
    d. distributors

18. In regard to wet barrel hydrants, which of the following is not true?
    a. Wet barrel hydrants have water inside of them right at street level.
    b. Wet barrel hydrants are usually installed in climates where freezing temperatures are a concern
    c. A typical modern wet barrel hydrant has more than one outlet.
    d. Each outlet is operated by its own operating stem and valve.

19. In regard to dry barrel hydrants, which of the following is not true?
    a. Dry barrel hydrants are typically found in freezing climates, but can be installed in nonfreezing climates.
    b. In a dry barrel hydrant, the water is held in the pipe serving the hydrant safely above the frost line.
    c. The hydrant valve is controlled by an operating nut at the top of the hydrant.
    d. The outlets on this type hydrant are threaded.

20. In the United States, it should be the goal to inspect and maintain fire hydrants:
    a. on an annual basis
    b. on a monthly basis
    c. on a semiannual basis
    d. every time a hydrant is used

21. As a pump operator, if your assignment was to secure a water supply, then to get the most water, you would choose the hydrant with which color bonnet?
    a. red
    b. light blue
    c. orange
    d. green

22. To determine the maximum flow of a hydrant and assess the available capacity of the water distribution system in the area requires that this type test be conducted:
    a. flush test
    b. flow test
    c. hammer test
    d. capacity test

23. What activity should take place before conducting flow testing?
    a. Hydrants should be flushed for a period of 24 hours.
    b. The water company must dig up the street in front of the hydrant.
    c. Notification of the test should be made to the water department.
    d. A hazmat team should be standing by the test the water.

24. Which of the following is not one of the three key items required for a triple combination pumper?
    a. master stream appliance
    b. pump
    c. fire hose
    d. water tank

25. An apparatus that has a water tank that holds up to 4,500 gallons of water is called:
    a. super pumper
    b. water tanker
    c. water tender
    d. auxiliary water unit

26. A basic water shuttle operation consists of three elements. Which of the following is not one?
    a. a dump site
    b. a fill site
    c. a turnaround zone
    d. the necessary fire apparatus

27. Put the water shuttle cycle in the proper order:
    1. The water tender is filled by an engine drafting from the static water source.
    2. A dumpsite by the fire with a portable tank initially filled by the water tender is established.
    3. A route to a nearby water source is established with the capacity to fill a water tender in a rapid fashion preferable at least 1,000 gpm.
    4. Once filled, the water tender returns to the dump site and fills the portable tank again.

    a. 3, 2, 1, 4
    b. 2, 3, 1, 4
    c. 3, 1, 2, 4
    d. 1, 2, 3, 4

28. To increase the amount of water being delivered by a shuttle operation, fire departments should do all of the following except:
    a. Use more portable tanks.
    b. Use multiple dump sites.
    c. Use more water tenders.
    d. Increase the speed of the apparatus.

29. A specialty hydrant that functions as an easy way to draft water from an untreated static water source is called:
    a. auxiliary hydrant
    b. drafting hydrant
    c. jet siphon
    d. cistern

30. A device that is typically used to move water from one portable tank to another in water shuttle operations in which water is pumped into a small 1½-in. intake and then rapidly, under high pressure, discharged into a large hard suction hose is called:
    a. auxiliary hydrant
    b. drafting hydrant
    c. jet siphon
    d. cistern

31. This is the main reason fire departments establish relay operations:
    a. unreliable water supplies
    b. friction loss in hose
    c. limited numbers of apparatus
    d. personnel limitations

32. When conducting a three-pumper relay operation, this pumper receives water from the source pumper and then pumps the water to a third pumper:
    a. supply pumper
    b. source pumper
    c. inline pumper
    d. attack pumper

33. The pumper that is last in the relay is called:
    a. supply pumper
    b. source pumper
    c. inline pumper
    d. attack pumper

34. The pumper located at the water source that pumps water to a second pumper is called:
    a. supply pumper
    b. source pumper
    c. inline pumper
    d. attack pumper

35. You have encountered a hydrant that is colored purple. What does this mean? Can you use it?
    a. This is a hydrant out of service. Do not use it.
    b. This is a hydrant in a new development. It is OK to use it.
    c. This is a reclaimed water hydrant. It is OK to use it.
    d. This hydrant also serves a standpipe system. It should only be used under extreme circumstances.

36. The classification of fire service hose is based on what factor?
    a. its construction
    b. its length
    c. the inherent friction loss
    d. its tactical uses

37. What type of hose is a rigid hose that is used by a pump to acquire water from static water sources?
    a. drafting hose
    b. supply hose
    c. attack hose
    d. wildland hose

38. What type of hose is a type of specialty attack hose designed to work well for combating vegetation fires and is not produced in diameters greater than 1½ in.?
    a. drafting hose
    b. supply hose
    c. attack hose
    d. wildland hose

39. What type of hose, sometimes referred to as hand line, must have a diameter large enough to move the required flow, but be small and lightweight enough to be easily moved around the fireground by firefighters?
    a. drafting hose
    b. supply hose
    c. attack hose
    d. wildland hose

40. What type of hose is designed to move larger volumes of water (and naturally these supply hoses have a larger diameter than attack hoses)?
    a. drafting hose
    b. supply hose
    c. attack hose
    d. wildland hose

41. Regular fire service pumps are not designed to operate efficiently at pressures above:
    a. 350 psi
    b. 400 psi
    c. 250 psi
    d. 150 psi

42. Regarding hose construction, what is the main purpose of the first jacket?
    a. Retain integrity while being heated.
    b. Resist the expansion of the waterway.
    c. Protect against foreign object penetration.
    d. Reduce friction loss.

43. Which of the following is commonly referred to as structure fire hose?
    a. extruded
    b. double jacketed
    c. single jacketed
    d. large diameter

44. Booster hose is commonly used on what type fires?
    a. nuisance fires
    b. car fires
    c. as an exposure line
    d. to wash bodily fluids down the sewer

45. This is a raised part of a coupling that allows a spanner wrench to grab onto it.
    a. cut
    b. lug
    c. bight
    d. tail

46. This sexless coupling is manufactured for use on all sizes of hose, with the most common sizes being 4-in. and 5-in.
    a. National Standard
    b. Victaulic
    c. Borscht
    d. Storz

47. Joining two male hose fittings together is made possible by using what type fitting?
    a. male reducer
    b. double male
    c. double female
    d. double expansion ring

48. Why do modern Storz fittings have locking mechanisms?
    a. Their weight requires a more substantial securing device.
    b. Firefighters can connect mismated couplings quickly and securely.
    c. They are susceptible to uncoupling if the hose twists violently.
    d. Firefighters often did not secure them properly.

49. A piece of equipment that water flows through that is portable in nature is known as _____.
    a. tool
    b. coupling
    c. appliance
    d. adaptor

50. This is a device where water enters through a single female inlet and then leaves through two male outlets.
    a. wye
    b. Siamese
    c. gate valve
    d. Water Thief

51. This device takes two female inlets and turns them into one male outlet.
    a. wye
    b. siamese
    c. gate valve
    d. portable hydrant

52. This tool was sometimes used to stop the flow of water to replace a burst section of hose.
    a. hose jacket
    b. hose clamp
    c. hose bridge
    d. hose roller

53. This is a hose tool that is put around a hose that has sprung a leak.
    a. hose jacket
    b. hose clamp
    c. hose bridge
    d. hose roller

54. You are assigned to secure fire hose to an aerial ladder during a ladder pipe operation. Which of the following would you use?
    a. hose clamp
    b. spanner
    c. gate clamp
    d. hose rope tool

55. Which of the following information regarding hose loading procedures is not true?
    a. When loading hose onto a fire apparatus make and attempt to reload the hose in the same manner in which is was deployed.
    b. Place the hose is such a manner that the folds are in different locations than before.
    c. Check the hose for serviceability, because it could have incurred serious damage during use.
    d. Check for obvious damage to couplings and the outer jacket.

56. What does "exercising the hose" refer to?
    a. stretching the hose in individual lengths to inspect it
    b. running water through the hose to check for leaks
    c. removing the hose from the apparatus and reloading it so that the folds occur at different positions
    d. stretching the hose into lengths of 300 feet and filling it with compressed air to check for coupling slippage

57. This is the pressure used to determine if fire hose is still suitable for use, and is the pressure the hose is tested to during the annual service test by the user.
    a. service test pressure
    b. proof/acceptance test pressure
    c. burst pressure
    d. operating/working pressure

58. Operating/working pressure should not exceed what percentage of the service test pressure?
    a. 50%
    b. 60%
    c. 75%
    d. 90%

59. When conducting hose testing, the first thing that needs to be accomplished is:
    a. a coupling tightness test
    b. a visual inspection of the hose
    c. identify the service test pressure
    d. inspect the hydrant to be used for the test

60. During hose testing, once the testing pressure has stabilized, the service test pressure shall be held for how many additional minutes?
    a. one minute
    b. three minutes
    c. five minutes
    d. ten minutes

61. You have come across a length of hose that is rolled in a straight roll with male coupling at the center of the roll. What is your action?
    a. It is out of service; tag it and place it in the proper area.
    b. It is in service; place it on the in-service rack.
    c. Unroll it and roll it back up with the female in the middle, then place it on the engine in a compartment.
    d. It is out of service; notify your officer.

62. After the deployment of all of the preconnected hose, at what point can the pump operator open the discharge outlet, supplying water to the line?
    a. as soon as all of the preconnected hose has cleared the hosebed
    b. when the preconnected hose has been properly flaked out
    c. when the attack team can be seen at the front door of the structure
    d. only after the call for water is made by the members manning the pulled preconnected attack line

63. When loading supply hose, it is sometimes necessary to take a short fold of hose to ensure the coupling does not flip over while paying out. What is this short fold of hose commonly called?

    a. Jersey flip
    b. forward fold
    c. Dutchman
    d. hose flap

64. What type of supply lay is used when an engine stops at the fire building and then lays hose to the nearest hydrant?

    a. modified lay
    b. forward lay
    c. reverse lay
    d. split lay

65. What type of supply lay is used when an engine spots a hydrant near the fire and then lays supply line toward the fire building?

    a. modified lay
    b. forward lay
    c. reverse lay
    d. split lay

66. Where would you place a pumper to best utilize all of the existing flow capacity in water main system?

    a. at the hydrant
    b. at the fire
    c. at draft
    d. in the center of a relay

67. What type of supply lay is used where two engines are used to complete one water supply line evolution from the hydrant to the building fire?

    a. modified lay
    b. forward lay
    c. reverse lay
    d. split lay

68. This type valve is attached to a hydrant to provide the option to pump at the hydrant after water flow is established through the initial supply line. What is it?

    a. gate valve
    b. Gleason valve
    c. LDH Humat four-way valve
    d. Carlin automatic hydrant valve

69. This valve allows the use of the other hydrant outlet without having to turn the hydrant off:
    a. gate valve
    b. Gleason valve
    c. LDH Humat four-way valve
    d. Carlin automatic hydrant valve

70. Suppose you are a pump operator. You have properly located a hydrant and positioned the apparatus. What is the next action to take in your water supply operation?
    a. Connect the supply hose to the hydrant discharge.
    b. Wait until the attack teams call for water to be supplied.
    c. Immediately check the hydrant for water by flushing it.
    d. Break and connect the supply line to the pumper's discharge.

CHAPTER 15 ANSWER KEY

Question #	Answer	Page #	Question #	Answer	Page #
1	d	388	31	b	401
2	c	388–389	32	c	401
3	b	389–390	33	d	401
4	a	390	34	b	401
5	c	390	35	c	402
6	d	390	36	d	403
7	c	390–391	37	a	405
8	c	391	38	d	406
9	a	391	39	c	406
10	c	391	40	b	406
11	c	392	41	c	404, 422
12	b	392	42	b	404
13	b	393	43	b	404
14	a	393	44	a	403
15	d	394	45	b	408
16	a	393	46	d	408
17	b	393	47	c	413
18	c	394	48	c	413
19	b	394	49	c	415
20	a	395	50	a	415–416
21	b	397	51	b	415–416
22	b	398	52	b	417–418
23	c	398	53	a	419
24	a	398	54	d	419
25	c	398	55	a	420–421
26	c	398–399	56	c	421
27	b	399	57	a	422
28	d	399	58	d	422
29	b	400	59	b	422
30	c	400	60	b	423

CHAPTER 15 ANSWER KEY *Continued*

Question #	Answer	Page #
61	b	423
62	d	427
63	c	431
64	c	431
65	b	431
66	a	433
67	d	435
68	c	436
69	a	438
70	c	439

Fire Streams

by Jay Comella with Jeff Shupe

1. Fuel loads today are largely comprised of which type materials?
    a. cellulosic
    b. inorganic
    c. block solids
    d. hydrocarbon

2. Most of America's fire service now considers this gallonage (gallons per minute, or gpm) figure as the minimum acceptable flow rate for interior structural fire attack for residential fires:
    a. 100 gpm
    b. 150 gpm
    c. 175 gpm
    d. 200 gpm

3. The minimum acceptable hand line flow rate for operations in commercial occupancies is:
    a. 150 gpm
    b. 200 gpm
    c. 250 gpm
    d. 300 gpm

4. The outcome of fireground operations depends on the outcome of the battle between two opponents: One is the water the engine company delivers, and what is the other?
    a. the size of the structure
    b. the orientation of the exposures
    c. the fire's heat release rate
    d. the personnel available

5. The flow at which the engine company can win the battle and kill fire is defined as:
   a. critical flow rate
   b. minimal flow rate
   c. nominal flow rate
   d. maximum flow rate

6. The single most important characteristic of a hose and nozzle system is:
   a. reliability of the water source
   b. water flow capability
   c. hose-coupling compatibility
   d. the size of the pumping apparatus

7. Water extinguishes fire primarily by what action?
   a. oxygen exclusion
   b. inhibition of ignition components
   c. reduction of temperature
   d. removal of fuel

8. When water turns to steam at 212°F, it absorbs an additional 970 British thermal units (Btu). This phenomenon is known as:
   a. positive heat balance
   b. sublimation
   c. ionization
   d. latent heat of vaporization

9. When liquid water turns to steam, it expands _____ times its original volume
   a. 1,000
   b. 1,200
   c. 1,500
   d. 1,700

10. Nozzle reaction is a function of which two factors?
    a. friction loss and flow rate
    b. flow rate and nozzle pressure
    c. nozzle pressure and nozzle discharge
    d. flow rate and nozzle discharge

11. What is the practical flow rate for 1¾-in. hose?
    a. 125 gpm
    b. 150 gpm
    c. 180 gpm
    d. 200 gpm

12. Nozzles do all of the following except?
    a. Reduce friction.
    b. Control flow.
    c. Create shape.
    d. Provide reach.

13. The rule of thumb for smooth bore tip orifice size is that it should be:
    a. one-quarter of the inside diameter of the hose
    b. one-half of the inside diameter of the hose
    c. the same size as the inside diameter of the hose
    d. twice the size as the inside diameter of the hose

14. This nozzle requires the least maintenance and has the longest service life:
    a. automatic fog nozzle
    b. constant gallonage fog nozzle
    c. smooth-bore nozzle
    d. adjustable gallonage nozzle

15. During interior fire attack, what type stream pattern should be used with a fog nozzle?
    a. wide fog
    b. narrow fog
    c. straight
    d. solid

16. This nozzle is designed to flow a specific gallonage when operated at the specific pressure for which it was designed:
    a. automatic fog nozzle
    b. constant gallonage fog nozzle
    c. solid bore nozzle
    d. adjustable gallonage nozzle

17. This nozzle uses a flow selection ring and requires an increased level of training for both the nozzle and pump operators:
    a. automatic fog nozzle
    b. constant gallonage fog nozzle
    c. solid bore nozzle
    d. adjustable gallonage nozzle

18. This nozzle produces a stream of reach and appearance consistent with 100 psi tip pressure regardless of the pressure actually coming into the base of the nozzle:
    a. automatic fog nozzle
    b. constant gallonage fog nozzle
    c. solid bore nozzle
    d. adjustable gallonage nozzle

19. A feature on fog nozzles that allows small pieces of debris to be discharged from the nozzle is known as the _____ feature.
    a. flush
    b. purge
    c. rinse
    d. sluice

20. The first step in planning a hose and nozzle system is:
    a. Determine the capacity of the water supply system.
    b. Establish a needed flow for the occupancy type in question.
    c. Determine where auxiliary appliance support is required.
    d. Establish a procedure for hoseline advancement.

21. What is the key to efficiently using a 2½-in. line?
    a. properly flaking the line
    b. adequate water supply
    c. proper staffing
    d. proper nozzle selection

22. You arrive at a two-story private dwelling with extensive involvement of the first floor and front porch. What hoseline will you choose to handle this situation?
    a. 2½ in.
    b. preconnected deck gun
    c. 1¾ in.
    d. 2 in.

23. In this method of fire attack, steam is a major factor in extinguishment, through cooling as well as smothering:
    a. direct attack
    b. modified direct attack
    c. indirect
    d. combination

24. In this attack, a fog stream is operated from a window or doorway and rotated in a clockwise direction to hit the ceiling, walls, and floor:
    a. direct attack
    b. modified direct attack
    c. indirect
    d. combination

25. This type of attack is used when the fire is relatively small and can be easily extinguished by applying water directly on the burning materials:
    a. direct attack
    b. modified direct attack
    c. indirect
    d. combination

26. This attack involves first applying the fire stream into the overhead area and then lowering the stream to quench the burning materials:
    a. direct attack
    b. modified direct attack
    c. indirect
    d. combination

27. Which is incorrect regarding the indirect attack?
    a. There must be no life hazard in the compartment.
    b. The atmosphere in the compartment must have reached at least 1000°F.
    c. Ventilation must be conducted as soon as steam creation is apparent.
    d. The stream must be injected into the compartment from the exterior.

28. The tactic(s) by which then fire department can most expediently improve the tenability of the fire compartment without compromising thermal balance is:
    a. aggressive interior direct attack and aggressive natural ventilation
    b. cautious interior direct attack and aggressive search operations
    c. aggressive interior direct attack and cautious search operations
    d. cautious interior direct attack and aggressive ventilation operations

29. When using this type fire stream prior to extinguishment, conditions in the fire area, especially in the fire compartment, worsen before they improve:
    a. solid stream
    b. fog stream
    c. straight stream
    d. broken stream

30. This type stream applied at a high rate of flow causes flame, smoke, and gas production to cease in the shortest amount of time, accomplished by the least amount of water:
    a. solid stream
    b. fog stream
    c. straight stream
    d. broken stream

31. What is the fire department's most valuable commodity regarding mission capability?
    a. apparatus
    b. training
    c. personnel
    d. funding

32. All doors through which a hoseline is stretched and advanced must be:
    a. forced by the ladder crew
    b. held open until the hoseline is charged
    c. marked as the attack route
    d. chocked in the open position

33. Who is responsible for removing kinks from a hoseline?
    a. All engine personnel are responsible.
    b. Probationary firefighters assigned to the attack team do this.
    c. It is everyone's responsibility.
    d. Ladder company personnel assigned to forcible entry do this.

34. The pump operator should be notified to charge the line only after:
    a. The fire is encountered.
    b. All personnel are positioned along the line.
    c. The line has been properly stretched and flaked out.
    d. Command has ordered the line charged.

35. Prior to entering the fire area, entrapped air is removed from the charged line. This is known as:
    a. purging the line
    b. raising the main
    c. exhausting the air
    d. bleeding the line

36. You have advanced an attack line to the door to the fire apartment door. Prior to opening the door, where should you position yourself?
    a. in front of the door
    b. with your back against the door
    c. to the side of the door
    d. across the hall from the door

37. In the modern fire environment, the prudent method of determining when to open the line is based on:
    a. when the officer orders it
    b. when the fire is encountered
    c. observation of the smoke condition
    d. the readiness of the ventilation and search teams

38. An indicator of highly heated smoke will be that the smoke will be:
    a. darker in color
    b. more pressurized
    c. more voluminous
    d. lower to the floor

39. You are advancing a line. There is an open compartment doorway that is issuing voluminous, pressurized, angry smoke. What action do you take?
    a. Retreat to a safe stairwell.
    b. Request that additional ventilation be conducted opposite the attack and at the roof level.
    c. Request that a backup line be stretched. Wait in a safe area until it arrives.
    d. Open the stream and sweep back and forth across the ceiling out front andahead of the nozzle team.

40. You have used the stream to sweep the superheated overhead of the compartment. In regard to stream application, what would you do next?
    a. Sweep the floor prior to advancing any further.
    b. Sweep the ceiling further into the fire area.
    c. Advance to the seat of the fire and apply water.
    d. Use the stream to break out a window to aid in ventilation efforts.

41. During the stretch, the nozzle operator must ensure there is a minimum of _____ feet of hose at or near the entrance to the fire area.
    a. 25
    b. 50
    c. 75
    d. 100

42. Which of the following are the best patterns of nozzle movement?
    a. sweeping side to side or clockwise rotation
    b. counterclockwise rotation or ceiling strike method
    c. sweeping side to side only or up and down
    d. clockwise rotation or up and down motion

43. Which of the following is incorrect regarding the nozzle team backup person?
    a. The backup person does the lion's share of the physical labor.
    b. The goal of the backup person is to make the job of the nozzle operator easier.
    c. The backup person should be on the same side as the nozzle operator.
    d. The backup person must be directly behind, but avoid physical contact with the nozzle operator.

44. What position must the door firefighter be in when operating as part of the nozzle team for a fire in a first-floor apartment?
    a. directly behind the backup person
    b. one or two lengths behind the backup person
    c. at the door to the apartment
    d. at the front of the building

45. Hydraulic ventilation is most effective with this type stream:
    a. straight streams from fog nozzles
    b. fog streams from solid bore nozzles
    c. fog streams from fog nozzles
    d. broken streams from special nozzles

46. When performing a nozzle vent with a fog nozzle, the first position of the stream pattern should be:
    a. straight stream
    b. narrow fog stream
    c. medium fog stream
    d. wide fog stream

47. Suppose a wall surface was being exposed to convected and radiated heat from a nearby fire. To keep the wall cooled below its ignition temperature, operations must concentrate on:
    a. placing a stream between the exposure and the fire source
    b. placing a stream directly onto the exposed surface
    c. placing a protective hoseline inside the exposed structure
    d. closing all windows on the exposed side of the structure

48. This is one of the first things firefighters should look for during overhaul:
    a. heavy loads
    b. tripping hazards
    c. laddering points
    d. structural damage

49. This special nozzle is a spinning nozzle with multiple orifices pointed in different directions, spraying water in all directions:
    a. ladder pipe
    b. cellar pipe
    c. Bresnan distributor
    d. piercing nozzle

50. This special nozzle has a directed stream that can be pointed in any direction through the use of a lever at floor level:
    a. monitor nozzle
    b. cellar pipe
    c. Bresnan distributor
    d. piercing nozzle

CHAPTER 16 ANSWER KEY

Question #	Answer	Page #	Question #	Answer	Page #
1	d	444	31	c	458
2	b	444	32	d	459
3	c	444	33	c	468
4	c	444	34	c	468
5	a	444	35	d	460
6	b	444	36	c	461
7	c	444	37	c	461
8	d	444–445	38	b	461
9	d	445	39	d	462
10	b	445	40	a	462
11	d	445	41	b	463
12	a	445–446	42	a	464
13	b	446	43	d	466
14	c	447	44	d	468
15	c	449	45	c	470
16	b	446	46	a	470
17	d	446	47	b	470
18	a	448	48	d	471
19	a	449	49	c	475
20	b	449–450	50	b	475
21	d	450			
22	a	451			
23	c	452			
24	d	453			
25	a	453			
26	b	454–455			
27	c	452–453			
28	a	454			
29	b	456			
30	a	457			

Firefighter Safety and Survival

by Anthony Avillo, Frank Ricci, and John Woron

1. The overriding concern of all fireground operations, firefighter training, and station activities is _____.

    a. organization
    b. discipline
    c. safety
    d. rules and regulations

2. What two factors is firefighter safety rooted in?

    a. discipline and attitude
    b. command and control
    c. training and awareness
    d. tradition and policies

3. This leading cause of death for firefighters accounts for close to half of the LODDs per year.

    a. apparatus and private vehicle accidents
    b. sudden cardiac death
    c. contact with live electrical equipment
    d. collapse of structures or structural components

4. During fireground operations, what should be in place to warn firefighters of dangerous building conditions?

    a. fireground barrier tape
    b. department safety bulletins
    c. a strong command presence
    d. emergency radio transmission protocols

5. Many times, firefighters become lost and disoriented because they are operating outside of the established attack plan. What is this fire service disease that cannot be tolerated at any level at any time called?

    a. improvising
    b. freelancing
    c. delegating
    d. inattention

6. This type of fireground injuries accounts for nearly half of all fireground injuries:
   a. strains and sprains
   b. slips and falls
   c. burns
   d. smoke inhalation

7. This is the first step in safeguarding personnel against exposures to infectious disease:
   a. documented training in infectious disease procedures
   b. recognition of the potential for exposure and contamination
   c. disclosure laws that protect firefighters and other emergency responders
   d. immediate decontamination of exposed or potentially exposed firefighters

8. Firefighter fatality and injury statistics exist for one reason. What is that?
   a. to recognize trends in casualties
   b. to identify unsafe equipment
   c. to pinpoint training needs
   d. so you do not become one

9. The number of firefighters killed in these types of buildings has been as much as four times the number that were killed in residential dwellings:
   a. commercial occupancies
   b. confined spaces
   c. vacant buildings
   d. high-rise buildings

10. This officer has the emergency authority to alter, suspend, or terminate any operation he or she deems hazardous to the operating personnel. On the fireground, this authority is on par with that of the incident commander.
    a. company officer
    b. accountability officer
    c. RIT officer
    d. safety officer

11. Many preventable injures occur at the fire station. How can these injuries best be remedied?
    a. through better supervision of firefighters
    b. by creation of a safety committee
    c. by proper and diligent house work
    d. by ensuring that all department property is properly maintained

12. Responding to and returning from an alarm is the second leading cause of firefighter line of duty death in the United States. Probably the biggest reason the casualty rate is so high is because:
    a. Driver safety policies and SOPs are not enforced.
    b. Civilian drivers often panic when fire apparatus approaches.
    c. Apparatus is often not properly maintained.
    d. Firefighters are exempt from speeding and red light regulations.

13. The equipment on the apparatus a firefighter is assigned to check on a daily basis will often be dictated by what?
    a. department maintenance schedules
    b. his or her seniority
    c. his or her riding assignment
    d. a directive from the company officer

14. If apparatus you are riding on comes in contact with live electrical lines, what is recommended action to take?
    a. Jump clear of the apparatus and do not touch the apparatus or the ground at the same time.
    b. Drive the apparatus out from under the fallen wires.
    c. Step off the apparatus lightly while maintaining contact with the handrail.
    d. Remain on the apparatus until the power is removed.

15. What is the first action to be taken when arriving at incidents at night or when visibility is obscured or expected to be obscured?
    a. Ensure that the apparatus slows before anyone dismounts.
    b. Utilize all illuminating devices and equipment.
    c. Notify dispatch of the condition and ask for assistance.
    d. Request that a light unit respond to the scene.

16. When on scene, especially an accident scene on a highway, what action is best in regard to apparatus positioning and roadway management?
    a. Close at least one lane in addition to the lane the accident is in.
    b. Conduct no operations until the police are on scene to manage traffic.
    c. Close all lanes to traffic in both directions.
    d. Utilize fire apparatus to shield all oncoming lanes. Place the hosebed closest to the traffic.

17. For traffic management for incidents on roadways, the distances for the advance warning and transition areas will differ depending on:
    a. magnitude of the incident
    b. number of emergency responders
    c. speed limit of the roadway
    d. weather conditions

18. If an emergency incident warrants the electrical power to a structure be turned off, what action should be taken?
    a. Firefighters should immediately locate the breakers and remove them.
    b. Fire personnel should pull the meter on the exterior of the structure.
    c. The incident commander should immediately request that the local utility turn off the power.
    d. Fire personnel should cut the power at the service connection to the structure.

19. Which is correct regarding natural gas?
    a. It should only be shut off by the gas company.
    b. When shutting it off, turn off the valve on the street side of the meter.
    c. When shutting it off, turn off the valve on the building side of the meter.
    d. The valve must be rotated a full turn to shut off the gas completely.

20. When encountering a gas-fed flame, what is the best action?
    a. Let the gas burn while protecting exposures.
    b. Extinguish the gas with a wide fog stream, then turn off the valve.
    c. Utilize foam to blanket the leaking gas.
    d. Evacuate the area until the utility company arrives.

21. You are about to enter a second-floor window in a residential structure. You have broken the window and cleared out the glass. Before entering, what action should you take next?
    a. Sweep the area below the window with a tool.
    b. Make sure you have secured a second means of egress.
    c. Sound the floor for stability.
    d. Request a hoseline to the area.

22. You have responded to a residential structure with a fire in the basement. You are with the attack team and are about to enter the basement to attack the fire. What action should you take before entering?
    a. Request horizontal ventilation of all basement windows prior to descending the stairs.
    b. Ensure that the search team enters the basement ahead of the attack line.
    c. Ensure that the line is charged before descending the stairs.
    d. Request an exterior stream to push the fire away from your descent.

23. When would be the only time a company deploys without direct orders?
    a. when on an additional alarm assignment and a Mayday is received
    b. when a firefighter is injured and the RIT is activated
    c. when a building collapse has occurred and firefighters can be seen in the rubble and are easily rescued
    d. when they are one of the first-arriving companies and their actions are being guided by scene assignment SOPs

24. You have been assigned to the rehabilitation division after working with your engine company at a second-alarm fire. The fire is now under control. Your company officer feels that you, along with the rest of the crew, are sufficiently rested and can go back to work helping pick up equipment. As a result, the rehab division supervisor is releasing your company from the rehab division. What action should you take next?
    a. Go to your apparatus and begin picking up your hose.
    b. Report to the command post.
    c. Use the portable radio to advise command of your available status.
    d. Prepare for release from the scene.

25. There are three crucial fireground operations that must go hand-in-hand to keep the operation as safe as possible. These are represented by the fireground coordination triangle. Which of the following is not one of these operations?
    a. fire attack
    b. fireground safety
    c. ventilation
    d. search and rescue

26. After entering a room via a window to conduct a search of that room, what is the most critical action to take to ensure firefighter safety?
    a. Close the door to the room.
    b. Request a hoseline to the stairwell outside the room.
    c. Search the room and the hallway outside the room for victims.
    d. Conduct additional ventilation of adjacent windows.

27. When searching a commercial building or a below-grade area such as a cellar or basement, what is the most important tool to utilize?
    a. thermal imaging camera
    b. set of forcible-entry irons
    c. lifeline
    d. portable radio

28. This is an absolute failure of the fireground at every level of the game:
    a. building collapse
    b. flashover
    c. burst hoseline
    d. emergency bailout

29. You have responded to a reported fire in a vacant building. Upon arrival, you are confronted with a marking on the building that looks like the image to the right. What does this mean?
    a. Operations may be conducted.
    b. This building is slated for demolition.
    c. Use caution when entering.
    d. Do not enter this building.

30. Under what circumstances would you transmit a Mayday transmission?
    a. localized collapse
    b. loss of water
    c. medical emergency
    d. report of an open shaft

31. When confronted with a Mayday situation, your survival is predicated on your ability to:
    a. Remain calm.
    b. Move to an upper floor of the building.
    c. Successfully utilize an emergency bailout.
    d. Breach barriers to escape in a timely and efficient manner.

32. Suppose you are lost in a building. To furnish the rapid intervention team (RIT) with a target to locate you, what acronym would guide your information to the RIT?
    a. SONAR
    b. LUNAR
    c. HELP
    d. MAYDAY

33. You have been assigned to the RIT at a high-rise fire. Where do you report and stage at this incident?
    a. at the command post in the street in front of the building
    b. at the command post in the lobby of the building
    c. at the operations division on the floor below the fire
    d. at the resource division two floors below the fire floor

34. You are inside a structure searching for fire and/or victims. You come across a door that opens toward you. Which of the following would you not be expecting to enter given the swing of the door?
    a. basement
    b. bedroom
    c. closet
    d. stairwell

35. When seeking to exit a structure, this is one room should never seek refuge in:
    a. bathroom
    b. garage
    c. kitchen
    d. bedroom

36. You are about to breach a wall. What is the first action that should be taken if you are to execute this option of escape?
    a. Check for an outlet on the wall.
    b. Sound a Mayday.
    c. Plunge the tool completely through wall at about 3½ ft off floor.
    d. Hit a stud sideways at the bottom to provide a wider opening.

37. You are conducting a "Detroit dive" maneuver to move from one area to another. You have encountered a plaster and lathe wall. What action must you take to ensure that the maneuver will be successful?
    a. Ensure that you completely remove the SCBA from your shoulders.
    b. Do not make the hole any more than 3 ft high, or it may cause a wall collapse.
    c. Ensure that the wall bay is completely cleared of building materials.
    d. Leave your tools in the original room until you have cleared the obstacle.

38. Which of the following is true regarding filter technology?
    a. Filters provide the minimum oxygen required for a firefighter to buy some exit time.
    b. To activate the filter, the firefighter must first remove his or her mask.
    c. Filters are rated to last up to 30 minutes.
    d. Among other gases, the filter provides protection against carbon monoxide.

39. The Reilly breathing technique uses what technique to conserve air?
    a. skipping a breath
    b. humming
    c. sharp breath utilization
    d. only breathing through the nose

40. At the very least, when should the RIT respond to an incident?

    a. with the first-alarm companies

    b. when requested by command

    c. with the striking of a second alarm

    d. upon confirmation of a working fire

41. Regarding rapid intervention, history has shown that a good majority of firefighters get themselves into trouble requiring assistance during what phase of the operation?

    a. response

    b. initial stages of the operation

    c. operations in elevated areas

    d. postcontrol operations

42. You and your partner are about to cut open a flat roof with a wood blade. Your partner is the saw operator. You are the guide person. Your partner has the saw revved up to full rpm and is about to sink it into the roof. What is your action?

    a. Tell your partner to bring the saw down to idle before sinking it into the roof.

    b. Move out of the circle of danger before the cutting starts.

    c. Put your hand on your partner's back and guide his or her cut.

    d. Tell your partner to not to cut until an examination hole is cut with an axe, so the saw may get a proper purchase in the roof.

43. Once on scene, the RIT should report to the command post for a briefing and consult and work closely with which officer?

    a. accountability officer

    b. rehab officer

    c. operations section chief

    d. incident commander

44. You are tracking a lost firefighter by the sound of his or her PASS device. Once the firefighter is found, what action should be taken to decrease the frenzy created by the situation?

    a. Get the firefighter on oxygen as soon as possible.

    b. Drag the firefighter to a more tenable area before assessing his or her condition.

    c. Request a hoseline to the area.

    d. Turn off the firefighter's PASS device.

45. Regarding air management, according to the NFPA 1404 Standard, "No firefighter will be allowed to operate with less than _____ tank of reserve air in an IDLH environment."

    a. ¼
    b. ⅓
    c. ½
    d. ¹⁄₁₀

46. When a downed firefighter is located, this simple action is critical and will greatly assist in a rapid removal of the firefighter.

    a. Attach the buddy breathing hose of the downed firefighter to your SCBA.
    b. Convert the downed firefighter's SCBA into a harness for easier removal.
    c. Request assistance from command and state your location.
    d. Request a hoseline to the area to protect against fire spread.

47. During a RIT operation to facilitate a quick exit and entrance path for egress, personnel, and resources, what action should be taken?

    a. Place a floodlight at the entrance to the building.
    b. Assign a firefighter to stand by at the entrance door to direct firefighters to the rescue point.
    c. Bring in a lifeline that can be secured to an object near the rescue point.
    d. Use power tools and widen the doorways between the rescue point and the egress door.

48. The drag rescue device (DRD) is designed to remove an incapacitated firefighter in which manner?

    a. vertically only
    b. horizontally only
    c. both vertically and horizontally
    d. via a lifeline for lifting only

49. When using the DRD, without employing a mechanical advantage, which of the following maneuvers provides the best pull power to remove an incapacitated firefighter?

    a. harness or belt drag
    b. two-firefighter Halligan tool rescue method
    c. narrow hallway maneuver
    d. one-firefighter Halligan hook rescue

50. What is the goal of the post-incident analysis?

    a. to identify training needs for the shift
    b. to develop a strategy to address issues in the incident evaluation
    c. to determine the effectiveness of SOPs
    d. to analyze after-action reports and dispatch tapes

CHAPTER 17 ANSWER KEY

Question #	Answer	Page #	Question #	Answer	Page #
1	c	480	31	a	532
2	c	480	32	b	532
3	b	482	33	c	537
4	d	483	34	b	521
5	b	483	35	a	539
6	a	484	36	b	539–540
7	b	486	37	c	540
8	d	487	38	d	541
9	c	517	39	b	541
10	d	494	40	d	542
11	c	500	41	b	542
12	a	501	42	c	543
13	c	502	43	a	543
14	d	502	44	d	546
15	b	503	45	a	546
16	a	504	46	b	547
17	c	504	47	c	547
18	c	502	48	b	548
19	b	509	49	a	549
20	a	509	50	b	554
21	b	483			
22	c	518			
23	a	515			
24	c	434			
25	a	516–517			
26	a	521			
27	c	522			
28	d	523			
29	d	528			
30	c	532			

Vehicle Fires

by Doug Leihbacher

18

1. Vehicle fires can occur when?
   a. when a vehicle is driven
   b. when the vehicle is parked
   c. as a result of an accident/collision that compromised vehicle systems
   d. all of the above

2. What is the primary hazard associated with vehicular fires that occur on high speed thorough fares?
   a. lack of water supply
   b. location of the vehicle or fire
   c. being struck by other high speed motorists
   d. time of day and weather conditions

3. When approaching a vehicle fire scence who should be responsible for sizing up the situation.
   a. officer
   b. apparatus driver
   c. police who arrive first
   d. all crew members

4. As a rule of thumb, the hose line should be advanced at a 45° angle from the sides of the vehicle?
   a. true
   b. false

5. What circumstances call for additional water supplies?
   a. vehicle fire impinging on a structure
   b. vehicle on fire is located on a divided highway
   c. limited amount of firefighters
   d. threat of the fire spreading beyond the vehicle

6. The recommended hoseline diameter for a vehicle fire is?
   a. 1 inch
   b. 2½ inch
   c. 1½ or ¾ inch
   d. 2 inch

7. Opening the hood (engine compartment) is necessary at vehicle fires for which reasons? Pick two.
   a. install battery cables
   b. complete fire extinguishment
   c. disconnect the battery cables
   d. check for fire

8. Fires in trucks and over-the-road vehicles present distinctive challenges to fire fighters because of:
   a. cargo, size, and type of vehicle
   b. hazardous materials, size, container type
   c. unknown materials
   d. greater amount of fuel in saddle tanks

9. What attack method should be used when a tanker truck is on fire?
   a. offense attack
   b. defensive attack
   c. combination defensive and offensive attack
   d. all of the above

10. Gasoline tankers can carry up to _____ gallons of fuel when full.
    a. 20,000
    b. 9,000
    c. 5,000
    d. 15,000

11. Liquefied petroleum gas (LPG) tank trucks can BLEVE in as little as?
    a. 7 minutes
    b. 3 minutes
    c. 5 minutes
    d. 10 minutes

12. Propane is 1.5 times heavier than air. When compressed, propane gas shrinks to 1/260 of its original volume.
    a. true
    b. false

13. At bus fires, _____ and _____ are the chief priorities.
    a. evacuation/extinguishment
    b. rescue/extinguishment
    c. scene safety/risk us benefit analysis
    d. evacuation/rescue

14. Some states allow school buses to be powered by propane or other alternative fuels.
    a. true
    b. false

15. The topmost priority at vehicle fires is what?
    a. life
    b. scene safety
    c. extinguishment
    d. exposure protection

16. The least common cause for vehicle fires is not what? Choose the correct answer or answers.
    a. design/construction/installation deficiency
    b. mechanical failure
    c. incendiary/suspicious
    d. misuse of heat

17. When fuel has been spilled at an accident scene, the use of flares _____:
    a. should not be utilized
    b. should be curtailed in favor of other warming devices
    c. should be used
    d. should be used only from a safe distance

18. The use of self-contained breathing apparatus (SCBA) at vehicle fires is _____ due to rubber and plastic gases contained in the smoke.
    a. required
    b. recommended
    c. suggested
    d. unnecessary because the fire is usually outside

19. _____ is an essential preliminary task when extinguishing a vehicle fire so it remains stationary once knockdown has taken place.
    a. Disconnecting battery cables
    b. Engaging the parking brake
    c. Securing the vehicle with cribbing or jacks
    d. Tire deflation

20. When attacking a vehicle fire, periodically aiming the stream down and banking it off the street allows water to bounce up and doesn't cool the under carriage and gas tank of the vehicle.
    a. true
    b. false

21. If the nozzle crew is crouching, the stream can be aimed upward toward the roof of the vehicle. This is preferable to prevent the stream from _____.
    a. not putting out the compartment fire
    b. breaking up
    c. advancing the fire condition in the vehicle
    d. going in one window and out the other side

22. The function of a ladder company at a vehicle fire on arrival is?
    a. search the vehicle
    b. pull back up hose lines
    c. provide scene safety for the first due engine company
    d. provide entry to engine compartment

23. The flammable range for gasoline in air is?
    a. 1.5% to 6%
    b. 1.5% to 6.5%
    c. 1% to 6%
    d. 0.5% to 6%

24. If a gasoline fuel tanker is involved, what class foam should be utilized?
    a. Class A
    b. Class B
    c. Class C
    d. Class D

25. Hybrid logos can be found on all of the locations except
    a. hatchback
    b. trunk
    c. doors
    d. front fenders
    e. front or rear bumpers

26. Hybrid and alternatively fueled vehicles are dangerous and extreamly hazardous under fire conditions. They should not be fought by conventional means.
    a. true
    b. false

27. If a fire involves the battery pack of a hybrid vehicle it should _____.
    a. not allowed to burn
    b. not be touched
    c. allowed to burn
    d. be extinguished and disassembled

28. Hybrids are _____ when stopped. It _____ possible to tell if it is running, which makes stabilization critical.
    a. silent/is
    b. loud/is
    c. humming/is
    d. silent/is not

29. Magnesium vehicle components that are on fire should be controlled by?
    a. Class D agents
    b. Class B foam
    c. water
    d. ABC extinguisher

30. Fires located within the engine compartment are difficult to extinguish until?
    a. The battery cables are disconnected.
    b. The head is opened.
    c. Hose streamers cool the hood area.
    d. Streams are directed through head lights.

31. Why is Class A foam not recommended for use when vehicles are on fire involving fuels?
    a. It is too expensive.
    b. Foam operations take too long to set up.
    c. The foam blanket becomes to large.
    d. It does not suppress flammable vapors long enough.

32. If piercing nozzles are unavailable, what tool can be used to penetrate the hood of the vehicle?
    a. flathead axe
    b. 6-ft pike pole
    c. Halligan
    d. bent tip applicator

33. Vehicle hoods all open from the front.
    a. true
    b. false

34. Disabling the battery
    a. prevents rekindle
    b. renders the air bags inoperable
    c. keeps the car from starting
    d. all of the above

35. By cutting the positive battery terminal cable it may produce a spark the could ignite _____ vapors.
    a. sulfure acid
    b. nitrogen
    c. argon
    d. hydrogen

36. If the vehicle's trunk needs to be opened the first action to take is?
    a. get the trunk key from the driver
    b. try the trunk release
    c. force entry
    d. wait for the towing company

37. If the vehicle registration, insurance card, or other documents are found within the vehicle they should be?
    a. turned over to the owner or police
    b. kept by the fire officer in charge
    c. left in the vehicle
    d. given to the two truck driver

38. Magnesium vehicle components are limited in cars today, and are not consumed quickly by fire that involve these parts.
    a. true
    b. false

39. In compressed natural gas (CNG) vehicles, the storage tanks are generally found _____.
    a. under the hood
    b. in the tank
    c. between the backseat and truck
    d. under the vehicle near the rear axle

40. Hybrid high voltage cables are what color usually?
    a. blue
    b. orange
    c. green
    d. red

41. Creating a bridge between the negative and positive battery cables is a consideration that is not important to fire fighters.
    a. true
    b. false

42. Over-the-road trucks are required by U.S. Department of _____ to display placards on all sides of the vehicle.
    a. Agriculture
    b. Homeland Security
    c. Commerce
    d. Transportation

43. The NFPA standard for recreational vehicles is
    a. NFPA 1192
    b. NFPA 25
    c. NFPA 1403
    d. NFPA 1500

44. A small pool of gasoline (2 sq. ft.) gives off approximately _____ KW of heat.
    a. 600 KW
    b. 50 KW
    c. 400 KW
    d. 250 KW

45. Whenever operating at taken incidents, efforts should be made to prevent spilled materials from entering _____ and _____.
    a. streams and ponds
    b. sewers and mapholes
    c. rivers and lakes
    d. brooks and sewers

46. Water can not be used to extinguish gasoline fires because water does not
    a. reduce the flashpoint
    b. control the burning fuel
    c. water can not handle the tremendous amount heat produced
    d. cool to a low enough temperature to prevent from generating flammable vapors

47. Supplemental equipment in addition to equipment needs for tanker fires includes master stream foam equipment, sand bags, dry chemical extinguishers, and _____.
    a. shovels
    b. sand trucks
    c. small excavator
    d. spill booms and absorbents

48. The placard number for gasoline tankers is _____.
    a. 1230
    b. 1203
    c. 1410
    d. 1650

49. Firefighters should stand _____ and uphill when at the scene of a gasoline spill.
    a. upwind
    b. downwind
    c. in front of
    d. behind

50. Firefighters are at greatest risk of injury when?
    a. stepping off the apparatus to attack a vehicle fire
    b. returning to pack tools and equipment after the fire is out
    c. during extinguishment
    d. during overhaul

CHAPTER 18 ANSWER KEY

Question #	Answer	Page #	Question #	Answer	Page #
1	d	560	31	d	568
2	c	561	32	c	569
3	d	561	33	b	570
4	b	566	34	d	571
5	a	564	35	d	571
6	c	565	36	b	573
7	b, c	569	37	a	573
8	a	577	38	b	573
9	d	580	39	c	575
10	b	579	40	b	575
11	c	584	41	b	575
12	b	584	42	d	578
13	d	586	43	a	578
14	a	587	44	c	580
15	b	561	45	b	579
16	b, c, d	559–561	46	d	579–580
17	b	562	47	c	580
18	a	563	48	b	579
19	c	564–565	49	a	579
20	b	565	50	b	562
21	d	565			
22	c	576			
23	a	580			
24	b	580			
25	e	575			
26	b	575			
27	c	575			
28	d	574			
29	a	576–574			
30	b	571			

Search and Rescue

by Mike Nasta with Joseph Alvarez

1. When using a Halligan tool for a search, which end should be used to "swing" toward the victim?

    a. adze end

    b. pointed end

    c. fork end

    d. shaft

2. Prior to entering a window to search an area for victims, firefighters should first:

    a. Ensure that the door to the room they are entering is closed.

    b. Sound the floor for stability.

    c. Gently sweep the area beneath the window for victims.

    d. Ensure that the power to the area has been shut down.

3. Before entering a building, every firefighter should make a mental note of this information:

    a. location of doors and windows

    b. location of the smoke

    c. location of hoselines

    d. location of the rapid intervention team

4. As you approach a structure on fire, you notice smoke being violently pushed out of the building, as if under extreme pressure. What should you be expecting here?

    a. backdraft

    b. wall collapse

    c. flashover

    d. BLEVE

5. Before opening any doors, you should always check for heat. How would you do this?
   a. by feeling the door knob with an ungloved hand
   b. by feeling the door from bottom to top
   c. by feeling the door from top to bottom
   d. by feeling the window nearest the door

6. You are searching a residential building and come across a door that opens away from you. What is likely to be behind the door?
   a. closet
   b. stairwell
   c. bedroom
   d. utility room

7. While searching, you and your partner come across a set of doors that have louvers built into them. What is likely to be behind that door?
   a. elevator room
   b. cellar
   c. utility closet
   d. office

8. What would be the floor search priorities when conducting a search of a five-story building with a fire on the second floor?
   a. 2, 3, 4, 5, 1
   b. 1, 2, 3, 4, 5
   c. 2, 5, 4, 3, 1
   d. 2, 3, 5, 4, 1

9. How much heat and smoke the room and floor directly above the fire receives depends on what factor?
   a. proximity to the stairwells
   b. the number of windows in the area
   c. the classification of building construction
   d. the fuel load in the fire area

10. The conditions on the top floor of a fire building depend on two factors. The first is the location and extent of the fire. What is the other factor?
    a. the number of stories in the building
    b. whether the floor has been properly ventilated
    c. how effective the attack operations are
    d. the square footage of the top floor

11. After entering the room to be searched, there are several areas that the searching firefighter must pay strict attention to. What would be the first area?

    a. behind doors

    b. beneath windows

    c. normal paths of egress

    d. center of the room

12. You are about to search a room and are in the process of forcing the door. You force the lock, but the door only opens partially. What action do you take?

    a. Feel the door for additional locks.

    b. Check behind the door for victims.

    c. Close the door and request a hoseline.

    d. Close the door and retreat—a backdraft is imminent.

13. This is a rapid search of high-priority areas usually conducted prior to control of the fire:

    a. primary search

    b. systematic search

    c. initial search

    d. precontrol search

14. This type search is slower and more methodical and is usually conducted after the fire has been controlled:

    a. postcontrol search

    b. perimeter search

    c. secondary search

    d. supplemental search

15. You are searching a residential structure and come across an adult victim on the second floor. Before removing the victim, what action should you take?

    a. Close all doors to prevent fire spread.

    b. Sweep the area for additional victims, possibly children.

    c. Request a hoseline to the area for protection.

    d. Ventilate the area to alleviate the heat conditions.

16. This search technique can be used if the smoke condition in the room is several feet off the floor; it enables the searcher to quickly scan the room and allows the area to be searched in a very short period of time:
    a. light scan search
    b. perimeter search
    c. lifeline search
    d. doorway search

17. You are engaged in a two-firefighter perimeter search. You have gone to the right, and your partner has gone to the left. You meet on the wall at the opposite side of the room. What action should be taken next?
    a. Crawl across the middle of the room back to the light by the entrance door.
    b. Briefly discuss what has been found thus far.
    c. Make your way to the next room to be searched.
    d. Pause momentarily and listen for any sounds of victims.

18. Suppose, during the two-firefighter perimeter search in question 17, after meeting on the other side of the room, the firefighters cannot see the light at the door. What action should be taken?
    a. Use a window to escape, and notify command.
    b. Use a lifeline to negotiate the middle of the room, and move back toward the light.
    c. Together, the two firefighters should follow one of the walls back to the doorway.
    d. Turn on an additional light to find the doorway.

19. You are searching in a hallway, and as you enter a room, you notice that the mattress has been folded in half. What action do you take?
    a. Search the room.
    b. Assume that there is a deceased fire victim wrapped in the mattress.
    c. Issue a Mayday.
    d. Move on to the next room. This room has been searched.

20. Suppose you are conducting a three-firefighter perimeter search. You are the third firefighter. The first and second firefighters have entered and begun searching along opposite walls. What is your assignment?
    a. Stand by and wait to direct the attack team.
    b. Conduct a light scan from the door with the light shining toward the middle of the room.
    c. Search down the middle of the room and meet both firefighters on the opposite wall.
    d. Close the door so as not to spread the fire.

21. Which of the following is not true regarding lifeline searches?
    a. This type of search should be used in any place with confusing or mazelike floor plans such as office cubicles.
    b. Perimeter search and light scan search techniques should not be used with lifeline searches.
    c. Indicator knots that can be tied in the rope at 20-ft intervals will allow the firefighter to determine how far he or she has penetrated the structure.
    d. The search rope can be tied to a nonmovable or substantial object outside the building or fire area, but not to a vehicle.

22. In large area buildings, what constant danger should firefighters be aware of?
    a. multiple victims
    b. flashover
    c. disorientation
    d. sprinkler runoff

23. You are conducting a lifeline search inside a large area structure. You come across an area where you must turn left. What should you do?
    a. Tie the rope off to keep it from dragging into unsearched areas.
    b. Pull the rope taut to minimize slack in the rope.
    c. Deploy a second rope into the new area.
    d. Leave the area and enter the structure via a different access point.

24. What can be done to expand the search while conducting a lifeline search?
    a. Leave a hand light at each turn.
    b. Attach tethers to the main line.
    c. Add personnel along the lifeline.
    d. Tie an additional lifeline to the end of the first lifeline.

25. During lifeline operations, the entry control officer records the entry time of each member and allows the team to operate for what portion of the SCBA cylinder's rated time?
    a. ¼
    b. ⅓
    c. ½
    d. ¾

26. Suppose you are the firefighter maintaining control of the rope bag and are ordered to exit the structure due to air cylinder duration limitations. Another crew is to replace your crew. What do you do with the bag of rope?

    a. Bring the bag out, but do not rebag the rope.
    b. Retrieve the rope and bag it on the way out.
    c. Leave the bag where it is and follow the rope out.
    d. Wait until the relief crew reaches your position, then follow the rope out.

27. How do thermal imaging cameras "see" through smoke?

    a. by means of an infrared lens
    b. by reading heat signatures
    c. by thermal coupling analysis
    d. by radiological metering

28. Because thermal imaging cameras can fail unexpectedly, what is recommended in the event that this failure occurs when using the camera while searching?

    a. Ensure that a second means of egress is established.
    b. Ensure that two cameras are brought into every building.
    c. Ensure that accountability tags are properly off at the command post before entering the building.
    d. Ensure that tried and proven search techniques are adhered to.

29. A safe and successful search combines the latest technology with:

    a. an established secondary means of egress
    b. the proper use of tools
    c. proper coordination of hoselines
    d. the best training

30. This SCBA maneuver is used when a firefighter encounters an entanglement hazard but is not yet entangled:

    a. inverted harness
    b. reduced profile
    c. swim method
    d. quick release

31. This SCBA maneuver will enable the firefighter to move through areas that would normally be too small to fit through:

    a. inverted harness
    b. reduced profile
    c. swim method
    d. quick release

32. This SCBA maneuver is used if the firefighter has already become entangled, and escape from the entanglement is necessary:
    a. inverted harness
    b. reduced profile
    c. swim method
    d. quick release

33. Regarding the blanket drag, which of the following is the best method?
    a. Drag feetfirst with head and shoulders on the ground, but supported.
    b. Drag headfirst with feet slightly off the ground.
    c. Drag headfirst with head and shoulders slightly off the ground.
    d. Drag headfirst with feet on the ground.

34. What is the goal of victim removal?
    a. to provide oxygen to the victim as soon as possible
    b. to ensure that firefighters are not injured in the process
    c. to not cause further injury to the victim
    d. to keep the fire away from the victim

35. Regarding victim removal, which of the following statements is not true?
    a. Victims should be removed to a clean area or removed to the outside of the structure as soon as possible.
    b. When removing a victim, make every effort to keep the victim's head as far from the floor as possible to prevent neck injuries.
    c. The practice of a firefighter removing his or her facepiece to give it to the victim must be strongly discouraged.
    d. Firefighters should try to remove victims to the exterior of a building by the path that gets the victim safely out of the IDLH (immediately dangerous to life and health) atmosphere as quickly as possible.

36. You have found a victim on the second floor of a two-story dwelling. The fire is on the first floor and is not yet under control. What action would you take regarding victim removal?
    a. Remove the victim via a second-floor window.
    b. Remove the victim via the interior stairs.
    c. Close the door to the room. When the fire is knocked down, remove the victim via the interior stairs.
    d. Protect the victim in place until a hoseline is available.

37. This firefighter carry can be performed by one firefighter on either a conscious or unconscious patient and is not recommended for a smoke-filled environment:
    a. seat carry
    b. firefighter carry
    c. blanket drag
    d. extremity carry

38. When removing victims, what part of the rescuer's body should be used to lift and move a victim?
    a. legs
    b. back
    c. arms
    d. legs and back

39. When using webbing to create a harness, what knot should be used to make a complete loop?
    a. bowline
    b. figure eight
    c. barrel knot
    d. water knot

40. What is the first step of the webbing drag?
    a. Loosen and unbuckle the waist strap.
    b. Roll the victim onto his or her back.
    c. Turn on the firefighter's SCBA purge valve.
    d. Slide the webbing under the victim's shoulder blades.

CHAPTER 19 ANSWER KEY

Question #	Answer	Page #	Question #	Answer	Page #
1	c	594	31	b	603
2	c	594	32	d	603–604
3	a	594	33	c	604
4	c	595	34	c	604
5	b	595	35	b	604
6	c	595	36	a	604
7	c	595	37	b	604–605
8	d	595	38	a	604
9	c	595	39	b	607
10	b	595	40	d	607
11	c	595			
12	b	595–596			
13	a	596			
14	c	596			
15	b	596			
16	a	596			
17	d	597			
18	c	597			
19	d	598			
20	b	598			
21	b	599			
22	c	598–599			
23	a	599			
24	b	600			
25	d	600			
26	c	600			
27	b	602			
28	d	602			
29	d	602			
30	c	602			

Basic Fire Attack

by Jerry Knapp with Christopher Flatley

1. A successful fire attack is the result of:
   a. an effective incident management system
   b. proper staffing at the firefighter ranks
   c. proper departmental planning for hazards
   d. the coordinated effort of multiple fire companies working together

2. This term refers to the development and implementation of a plan, incorporating specific goals, to bring a fire under control:
   a. strategy
   b. command
   c. tactics
   d. control

3. This term refers to the specific actions taken to accomplish the identified goals:
   a. strategy
   b. command
   c. tactics
   d. tasks

4. Stating the height of the building in the on-scene report will provide information regarding the:
   a. amount of hose required
   b. general knowledge of the typical hazards
   c. type of ground or aerial ladders needed
   d. extent of the potential collapse zone

5. This information paints the picture for incoming units, so they can begin to consider what priorities, actions, and tools will be necessary for the call:
   a. on-scene report
   b. prefire plan
   c. dispatch report
   d. standard operating procedures

To answer questions 6–8, consider the following diagram:

6. If you were standing in front of the building, what side of the building would you be on?

    a. side A

    b. side B

    c. side C

    d. side D

7. What would the building to the left of the fire building be designated?

    a. exposure A1

    b. exposure B1

    c. exposure B

    d. exposure D

8. If you were being assigned to search the building where the smoke is issuing from the cockloft (on the extreme right), what exposure would you be entering?

    a. exposure D1

    b. exposure B2

    c. exposure D3

    d. exposure D2

9. This is the ongoing process of evaluating the situation before arrival, upon arrival, and during the incident:

    a. prefire planning

    b. command evaluation

    c. size-up

    d. progress report

10. What is the first step in resolving any emergency situation?
    a. Familiarize yourself with the building.
    b. Conduct a size-up.
    c. Position the hoseline.
    d. Secure a water supply.

11. On a structural fire response, if you heard the term "new construction" in the size-up, what should you be thinking?
    a. Heavy fire load is present.
    b. There are lightweight components.
    c. No masonry is present.
    d. A sprinkler system is in the building.

12. In regard to construction hazards, what are the firefighter's best survival tools?
    a. prefire intelligence and planning
    b. accountability tag systems
    c. thermal imaging cameras
    d. forcible-entry tools

13. At this point in your career, many of the strategic decisions will not be made by you. Your best action on the fireground at this time would be to:
    a. Give only tactical input.
    b. Observe everything and say nothing.
    c. Pass on your size-up concerns to your superior.
    d. Control your anxiety and follow orders.

14. When does size-up begin?
    a. when you arrive on scene
    b. at the receipt of alarm
    c. when you are entering the fire building
    d. during response

15. Who is responsible for conducting continuous size-up?
    a. the incident commander
    b. company and chief officers
    c. the safety officer
    d. all personnel on the fireground

16. This strategy is used when an aggressive interior search or attack can save lives or property:
    a. offensive
    b. defensive
    c. indirect
    d. offensive/defensive

17. This strategy is used, for example, when a building is unsafe due to a fire load on lightweight construction components:
    a. offensive
    b. defensive
    c. defensive/offensive
    d. offensive/defensive

18. This strategy is used when there are hazards that must be controlled before firefighters make entry to a building or area:
    a. offensive
    b. defensive/offensive
    c. defensive
    d. offensive/defensive

19. The strategy is used when the fire is progressing faster than our ability to put water on it, where we may need to switch to a defensive position:
    a. offensive
    b. defensive/offensive
    c. defensive
    d. offensive/defensive

20. What can replace the flat-head axe as a tool to drive a Halligan tool?
    a. sledgehammer
    b. pick-head axe
    c. pike pole
    d. ball peen hammer

21. There are two important safety factors to consider when conducting forcible-entry operations. Pick two of the following:

    1. Consider forcible exit.
    2. Ensure that a hoseline is always in place when forcing entry.
    3. Always control the door.
    4. Carry additional tools.

    a. 1 & 2
    b. 2 & 3
    c. 1 & 3
    d. 3 & 4

22. Our first priority must always be to:
    a. Protect exposures.
    b. Save the occupants of the building.
    c. Extinguish the fire.
    d. Control loss.

23. What is the primary purpose of the first hoseline?
    a. to extinguish the fire at its seat
    b. to confine the fire to the area of origin
    c. to protect both interior and exterior exposures
    d. to protect the means of egress and the rescue operation

24. The preferred means of victim removal is via:
    a. aerial ladder
    b. fire escape
    c. interior stairs
    d. adjoining building

25. Fire growth in compartment fires (inside a building) is often determined by:
    a. the building's construction
    b. the amount of oxygen available
    c. the degree of forcible entry required
    d. the layout of the room

26. What should the nozzle operator have on hand to begin the fire attack?
    a. just the nozzle
    b. forcible-entry tools
    c. the nozzle and the first 50 ft of hose
    d. the nozzle and the first 100 ft of hose

27. After removing his or her assigned equipment and heading toward the building, why must the nozzle operator wait for a few seconds before continuing with his or her assigned task?
    a. to allow the backup firefighter to shoulder his or her assigned equipment
    b. to determine the best place to attack from
    c. to more accurately size-up the structure
    d. to wait for the officer to don his or her SCBA

28. You are preparing to call for the line to be charged. What must you do at this point?
    a. Feel the door for heat.
    b. Make sure that all personnel are in place and ready.
    c. Call for the door to be forced.
    d. Ensure that the hoseline is properly flaked out.

29. The line has just been charged and is solid and fully charged with water. What is your next action?
    a. Tell the forcible-entry team to "take the door."
    b. Have the ventilation team to "take the windows."
    c. Open the line to check flow, pressure, and pattern.
    d. Ensure that the bypass valve on your SCBA is open.

30. Which of these reduces the possibility of flashover and backdraft?
    a. hose streams set on wide fog
    b. holding off on ventilation until the fire is found
    c. good, timely ventilation operations
    d. properly worn personal protective equipment

31. What is the primary responsibility of the backup firefighter?
    a. to relieve the nozzle operator
    b. to flake excess hose outside the operating area
    c. to relieve nozzle reaction from the nozzle operator
    d. to assist the nozzle operator in holding the nozzle steady

32. If you are assigned to this position, your job is to make sure the preconnect bed is clear of hose and to remove kinks from the attack line:
    a. pump operator
    b. nozzle operator
    c. backup firefighter
    d. door firefighter

33. Water hammer can be prevented by:
    a. maintaining hydrant systems
    b. always backing up your attack lines
    c. properly flaking out hoselines
    d. opening and closing nozzles slowly

34. In a well hole stretch, one length of hose will, on the average, reach from the base of the stairs to which floor?
    a. third
    b. fourth
    c. fifth
    d. sixth

35. You are in a high-rise fire, and you are about to commence the attack. The door to the hallway in front of you is relatively hot. There are people in the stairwell above. What do you do?
    a. Use water on the door to cool it before opening it.
    b. Hold the door closed until the stairs are clear.
    c. Ensure that you have water, open the door, and commence the attack.
    d. Switch your attack to another stairwell.

36. What is the most common type of fire attack used?
    a. direct
    b. modified direct
    c. indirect
    d. combination

37. The phenomenon where the hottest layer of air is in the highest part of the room, near the ceiling, and the coolest layer of air is at the lowest level, near the floor, is called:
    a. thermal radioactive feedback
    b. thermal layering
    c. flashover
    d. heat balance

38. Why is the use of fog streams or spray patterns for an aggressive interior fire attack dangerous?
    a. They create massive amounts of steam.
    b. There is more nozzle pressure on these streams.
    c. They are difficult to control in the interior of a building.
    d. Fog stream results are unpredictable.

39. Experienced firefighters often describe this operation as the key to success for any aggressive interior fire attack:
    a. reliable water supply
    b. continuous line advancement
    c. search and rescue
    d. ventilation

40. During ventilation, windows and doors are often used for this type of vent operation:
    a. indirect
    b. horizontal
    c. vertical
    d. direct

41. What is the major danger when using positive pressure ventilation (PPV) simultaneously with fire attack?
    a. The additional oxygen may cause an explosion.
    b. Fans will get in the way of attack teams.
    c. Fire may be pushed into voids and/or at trapped victims.
    d. More smoke may be created by the draft.

42. What is the first step in confining and limiting the spread of fire and ultimately extinguishing the fire?
    a. size-up
    b. protection of exposures
    c. closing doors in the fire building
    d. roof ventilation

43. A building that, left unprotected, will ignite by radiant, convected, or conducted heat is known as a(n):
    a. exterior exposure
    b. target hazard
    c. fire break
    d. perimeter hazard

44. To properly protect an exposure, water should be applied:
    a. directly on the walls of the exposure
    b. in between the fire building and the exposure
    c. in the windows of the fire building
    d. in the windows of the exposure

45. During overhaul, where floors lie on beams or where boards are nailed together, what action can be taken to stop smoldering and cool the area?
    a. Drive the hose stream between the boards.
    b. Monitor the area with a thermal imaging camera.
    c. Dismantle the area and remove it to the outside.
    d. Use a positive pressure fan to cool the area.

46. When picking up after the fire, what action should be taken before packing hose back on the engine?
    a. Turn the hose inside out to check the inner jacket.
    b. Spray grease or a suitable lubricant on the couplings.
    c. Make sure all hose is disconnected at the couplings on the hosebed.
    d. Inspect the hose for damage.

47. What is the purpose of the after action review (AAR) conducted before leaving the scene?
    a. to ensure that all equipment is accounted for
    b. to critique members on areas where improvement is needed
    c. to capture any lessons learned from the fire
    d. to shed light on meritorious acts

48. If you were responding to a fire where wooden I-beams were used, what would be your main concern?
    a. The drywall attached to its underside should protect it for approximately an hour.
    b. Collapse of the structure must be considered.
    c. Overhaul of the area should not be terribly difficult.
    d. The combustible components of the wooden I-beam will add fuel the existing fire load.

49. A good rule of thumb regarding required hoselines in a residential structure is one line per ___ room(s):
    a. one
    b. two
    c. three
    d. four

50. You are the engine pump operator supplying water to a single 1¾-in. attack line being operated in a private dwelling fire. You do not yet have water supply and are supplying the line from a 500-gallon booster tank. Approximately how much time do you have to secure a sustained water supply?
    a. 1 minute
    b. 2 minutes
    c. 3 minutes
    d. 5 minutes

51. What are most structure fires are controlled by?
    a. amount of air they have available
    b. size of the affected space
    c. height of the ceilings
    d. compartmentation in the structure

52. The heart of the fire attack is:
    a. the removal of victims
    b. an effective size-up
    c. the stretching and operating of the first hoseline
    d. adequate water supply

53. What is the key to success for a fire attack, which will result in rapid fire suppression and saving significant property?
    a. an adequate water supply
    b. an effective command organization
    c. rapid and effective ventilation
    d. a rapid primary search

54. What is the key to safe and successful response to basement fires in regard to fire attack?
    a. Plan, as a first action, to use outside streams to knock the fire down.
    b. Ensure that there are several ways out of the basement.
    c. Attack from below or at least from the same level.
    d. Ensure that backup lines of large-diameter hose are ready to go.

55. What is the best way to attack a basement fire in a residential dwelling?
    a. Use an interior attack via the clamshell doors on the exterior.
    b. Use streams through the exterior windows.
    c. Use an interior attack by descending the interior basement stairs.
    d. From the interior basement door on the first floor, discharge water directly down the interior basement stairwell.

56. What tool(s) would you use to force clamshell doors?
    a. axe and Halligan
    b. hydraulic forcible-entry tool (rabbit tool)
    c. metal-cutting saw
    d. K-tool with bent end key tool

57. Why is a basement the worst place for a fire in the building?
    a. Locked doors eliminate entry possibilities.
    b. The entire building is threatened.
    c. Illegal occupancies may exist.
    d. Ventilation is not available.

58. You have responded to a fire in a wood frame garden apartment complex. Conditions and layout prevent your attack team from accessing the fire apartment. What alternative can you take?
    a. Knock the fire down from the exterior.
    b. Look for an alternate access point at the rear.
    c. Place a positive pressure fan behind the attack team.
    d. Breach the wall of an adjacent unit, and apply water though the hole.

59. Regarding taxpayer fires that have extended to the cockloft, what is the best action?
    a. Stretch a line to the top floor of the fire building.
    b. Cut a large primary vent hole directly above the fire.
    c. Stretch a line to the top floor of the downwind exposure.
    d. Begin a trench cut near the leeward exposure.

60. In a one-story taxpayer of noncombustible construction, what is the preferred ventilation tactic?
    a. Cut the roof as directly over the fire, as far as it is safe to do so.
    b. Ventilate horizontally at the windows and doors.
    c. Ventilate horizontally at the rear by breaching the block wall.
    d. Do not ventilate this type fire.

61. In big box stores, these provide an excellent horizontal ventilation opportunity:
    a. show windows
    b. loading dock doors
    c. deck guns from engine companies
    d. positive pressure fans

62. What is most required when searching in an office building?
    a. charged hoseline
    b. thermal imaging camera
    c. forcible-entry tools
    d. lifeline

63. What actions should be taken at a fire in a fast-food restaurant where the fire has originated in or has extended to the area above the drop ceiling?
    a. Evacuate the structure, and operate defensively.
    b. Use large-caliber hand lines.
    c. Pull the ceilings from a protected area.
    d. Cut a hole in the roof directly above the fire.

64. When confronted with a fire in a vacant building, firefighters on the scene must assume that:
    a. The fire was set by an arsonist.
    b. The fire will be small because usually there are no furnishings inside.
    c. Water supply in the area will be inadequate.
    d. The structure is compromised.

65. For an "odor of smoke in the building" response, what is your most important consideration?
    a. Who turned in the alarm?
    b. What are the exposures?
    c. Is there fire in the voids?
    d. Is the smell actually in the building?

66. When responding to an outside fire in a rubbish container, water should be directed first:
    a. directly on the burning materials from a few feet away
    b. in an injection fashion deep inside the container
    c. from a distance toward the burning materials
    d. indirectly from above in the form of a fog pattern

67. What is the most common hazard in outbuilding fires?
    a. lack of water supply
    b. hazardous materials
    c. forcible entry
    d. wildland interface

68. What is the key to extinguishing small brush fires?
    a. Determine the direction in which they are heading.
    b. Determine what is burning.
    c. Determine which exposures need protection.
    d. Determine if there are any exotic hazards in the brush.

69. This portion of a wildland fire has the greatest rate of spread:
    a. crown
    b. head
    c. flank
    d. fingers

70. This type of wildland fire spreads quickly with a well-defined head:
    a. crowning
    b. backing
    c. running
    d. creeping

71. In a wildland fire, any sudden acceleration in the rate of spread for a short period of time is called a _____
    a. blow-up
    b. flare-up
    c. slopover
    d. control fault

72. These are natural or constructed barriers used to contain the fire:
    a. control lines
    b. anchor points
    c. fire lines
    d. fire barriers

73. This is the most common hand tool used in wildland fires
    a. torch or flare
    b. pick shovel
    c. Pulaski
    d. gizmo

74. Name two challenges for firefighters when operating at stacked materials fires:

> 1. disorientation
> 2. water application
> 3. postcontrol overhaul
> 4. collapse hazards

   a. 1 & 3
   b. 2 & 4
   c. 2 & 3
   d. 1 & 4

75. What can be used to increase the effectiveness of water on stacked materials fires?
   a. inerting ingredients
   b. carbon dioxide (CO_2)
   c. Class A foam
   d. sand

CHAPTER 20 ANSWER KEY

Question #	Answer	Page #	Question #	Answer	Page #
1	d	612	31	c	629–630
2	a	612	32	d	630
3	c	612	33	d	633
4	c	612	34	c	635
5	a	612	35	b	637
6	a	614	36	b	638
7	c	614	37	b	639
8	d	614	38	a	640
9	c	614	39	d	641
10	b	615	40	b	642
11	b	616	41	c	640
12	a	616	42	b	643
13	d	616	43	a	643
14	b	616–617	44	a	644
15	d	617	45	a	644
16	a	617	46	d	645
17	b	618	47	c	646
18	b	618–619	48	b	647–651
19	d	619	49	b	647
20	a	620	50	b	648
21	c	621	51	a	649
22	b	621	52	c	649
23	d	622	53	c	641
24	c	623	54	c	650
25	b	625	55	a	652
26	c	627	56	c	652
27	a	628	57	b	653
28	d	629	58	d	654
29	c	628–629	59	b	655
30	c	629	60	b	657

CHAPTER 20 ANSWER KEY *Continued*

Question #	Answer	Page #
61	b	658
62	d	659
63	a	660
64	d	661
65	c	662
66	c	664
67	b	665
68	a	666
69	b	666
70	c	666
71	b	666
72	a	667
73	c	667
74	b	668
75	c	668

Salvage and Overhaul
by Jeff Shupe

1. Salvage can be simply defined as:
    a. the reduction of water damage
    b. determining that the fire is completely out
    c. putting out the fire
    d. the art of saving property

2. Salvage is also known as:
    a. postoverhaul
    b. loss control
    c. water elimination
    d. loss prevention

3. This type of damage is caused by the combustion process, where there is physical destruction from flame, heat, or smoke:
    a. initial damage
    b. actual damage
    c. direct damage
    d. serial damage

4. This type of damage includes all the other damage caused by the suppression efforts and other activities:
    a. collateral damage
    b. extensive damage
    c. supplemental damage
    d. indirect damage

5. Salvage falls under which of the incident priorities of firefighting?
    a. firefighter safety
    b. life safety
    c. incident stabilization
    d. property conservation

6. Salvage begins with what other fireground operation?
   a. fire attack
   b. water supply
   c. primary search
   d. overhaul

7. When beginning salvage operations, the point where the initial salvage efforts should be conducted is determined by:
   a. the area with the least fire damage
   b. where fire operations ended
   c. the area that will have the greatest impact on saving property
   d. the lowest area of the building

8. Salvage covers can be most efficiently utilized when:
   a. The contents of the room are grouped together for covering.
   b. They are used to route water rather than cover items.
   c. They are clean and undamaged.
   d. They are laid on stairs prior to entry.

9. In what two places will salvage operations generally begin?

   1. fire floor
   2. floor below the fire
   3. the lowest point in the building
   4. on the exterior of the building

   a. 1 & 2
   b. 2 & 3
   c. 1 & 3
   d. 3 & 4

10. This is probably the most common salvage tool:
    a. mop
    b. squeegee
    c. salvage cover
    d. water vacuum

11. Put the following salvage cover maintenance protocol in order:

    1. Hang the salvage cover to dry.
    2. Check for any holes or tears.
    3. Wash the salvage cover.
    4. Fold the salvage cover.

    a. 1, 2, 3, 4
    b. 2, 3, 1, 4
    c. 3, 1, 2, 4
    d. 3, 2, 1, 4

12. Because of efficiency and disposability, canvas salvage covers are being replaced by:

    a. vinyl covers
    b. polyethylene plastic
    c. absorbent socks
    d. recycled paper sheets

13. Salvage covers are primarily used for this reason:

    a. to hold water
    b. to divert water to areas of proper removal
    c. to protect flooring and stairways
    d. to protect furniture and other valuables

14. This is a trough made of salvage covers, pike poles, and ladders to catch dripping water and divert it away:

    a. water chute
    b. water trench
    c. catchall
    d. carryall

15. If you build up or roll the sides of a salvage cover, once again using pike poles, what have you constructed?

    a. water chute
    b. water trench
    c. catchall
    d. carryall

16. You are doing salvage in a multistory building and need to drain water from an upper floor. It is not possible, due to the layout to create chutes or catchalls. What can you do?
    a. Punch a hole in the floor.
    b. Remove the toilet and sweep the water to the opened drain.
    c. Use the stairwell to remove the water.
    d. Open the windows and allow the floors to drain naturally.

17. Using pumps and water vacuums inside buildings to remove water is known as:
    a. dewatering
    b. postcontrol removal
    c. vac-pumping
    d. deflooding

18. Before entering an area with standing water, what action must be taken?
    a. A plan must be established on how to remove the water.
    b. The area must be well lit.
    c. Electricity must be shut down.
    d. Personnel should be relieved by fresh crews.

19. Following orders, you have stopped the flow from a sprinkler main supply by shutting down the OS&Y valve. What action should be taken next?
    a. Begin to bring attack hose out of the building.
    b. Close the water motor gong valve to silence the alarm.
    c. Open the main drain valve and drain the system.
    d. Shut the hydrant supplying the system.

20. You are covering a roof hole after a fire operation. One of the firefighters has brought Visqueen polyethylene sheeting to the roof. What do you tell him to do with it?
    a. Take it back. Bring a canvas tarp.
    b. Bring it to the top floor and cover the hole from the inside. Then put a canvas tarp over it.
    c. Place it over the hole and staple it in place.
    d. Place it over the hole, but make sure it is at least double-layered to prevent it from tearing.

21. When securing a building, what should firefighters do if the building has sustained damage beyond repair and will need to be demolished?
    a. Use as much water as needed to ensure rekindling will not happen.
    b. Remain on the scene until the building is demolished and removed.
    c. Establish a safe perimeter around the building with barrier tape.
    d. Leave the building in the custody of the owner.

22. When do overhaul operations begin?
    a. upon arrival of the fire department on scene
    b. once the fire is extinguished or under control
    c. with the application of the first attack line
    d. when the roof is opened

23. Before overhaul begins, what must be accomplished?
    a. The safety of the building must be determined.
    b. Excess water must be removed from the building.
    c. A roll call must be completed.
    d. Firefighters must be removed from the building.

24. Where does overhaul usually begin?
    a. at the fire's location within the building
    b. at the main entrance point, working inward
    c. on the floor above the fire, working downward
    d. on the roof, working downward

25. These two actions will help keep everyone aware of the fire's activities—its location, what it is doing, and where it might be going:

1.	visual investigations
2.	opening up of all walls
3.	ventilation operations
4.	radio reports

    a. 1 & 3
    b. 2 & 3
    c. 3 & 4
    d. 1 & 4

26. If, during overhaul, clues to the fire's origin are found, what should a firefighter do?
    a. Remove the material and finish overhauling.
    b. Dig around the area until a clear area is found.
    c. Preserve the evidence.
    d. Take a picture of the area before overhauling.

27. When should attack lines be removed from the structure?
    a. when smoke conditions have subsided
    b. when thermal imaging equipment shows no signs of heat
    c. when salvage operations are underway
    d. when the incident commander has approved their removal

28. What would be the best way to detect hidden fire?
    a. Use a thermal imaging camera to detect heat.
    b. Use your senses of sight and sound.
    c. Open walls that are suspect.
    d. Use the back of your hand to feel suspected surfaces.

29. Suppose you are opening a wall in a balloon frame building. As you open it up, fire issues from the wall space behind the hole. What action would you take?
    a. Continue to open it until the wall is completely down.
    b. Use a hose stream to direct the water downward into the space.
    c. Use a hose stream to direct the water upward into the space.
    d. Use a hose stream to direct the water upward and downward into the space.

30. You are overhauling a room and come across some small valuables that are in the open. Which is the best action?
    a. Lift them off the ground to prevent their being damaged by water runoff.
    b. Put them in a drawer or closet to protect them.
    c. Hand them over to the police.
    d. Put them in a box for safekeeping until they are returned to the owner.

CHAPTER 21 ANSWER KEY

Question #	Answer	Page #
1	d	673
2	b	673–674
3	c	674
4	d	674
5	d	674
6	a	674
7	c	675
8	a	674
9	a	675
10	c	676
11	c	676–677
12	b	677
13	d	677
14	a	677
15	c	677
16	b	678
17	a	678
18	c	678
19	c	679
20	d	677
21	c	680
22	b	680–681
23	b	677
24	a	680–681
25	d	682
26	c	682
27	d	686
28	a	686
29	d	684
30	d	676

Emergency Medical Response

by Mike McEvoy with Victor Stagnaro

1. After assessing your patient, you determine more advanced care is required. You should:

    a. Lay the patient on the ground for further examination.

    b. Never attempt to do more than your training has covered

    c. Place the patient into the care of the local law enforcement for them to seek additional help.

    d. Go ahead and start to treat the patient even if the steps you are taking are above your scope of knowledge.

2. The Health Insurance Portability and Accountability Act (HIPAA) was enacted in 1996. In general, the law states:

    a. Patients with communicable diseases are entitled to health care.

    b. A patient's medical history is public knowledge.

    c. A medical professional can deny treating a patient who does not have health insurance.

    d. An individual's health information is private and can only be disclosed under three circumstances.

3. When evaluating a patient, what is the *first* thing to do to determine the patient's level of consciousness?

    a. Open the patient's mouth and look for an obstruction.

    b. Call loudly to the patient while tapping or firmly shaking his or her shoulders.

    c. Open each eyelid to look for pupil reaction.

    d. Check for a radial pulse in each wrist.

4. If you suspect a spinal injury, how should you initially attempt to open the patient's airway?

    a. Don't move the patient, or you might paralyze him or her.

    b. Use the head-tilt, chin-lift method.

    c. Use the jaw thrust method to open the airway.

    d. Place your fingers on the back of the neck for support.

5. When checking for a pulse in a conscious adult, where is the best pulse on the body to do this?
    a. radial pulse in the wrist
    b. carotid pulse in the neck
    c. brachial pulse in the upper arm
    d. femoral pulse in the groin

6. Once the patient's airway, breathing, circulation (ABC) have been established, the next assessment is:
    a. rapid primary patient assessment
    b. focused primary patient assessment
    c. rapid secondary patient assessment
    d. focused secondary patient assessment

7. During the rapid primary patient assessment, your patient is checked from the head and neck, moving all the way down to the bottoms of the feet and toes. What are you looking for?
    a. medical alert bracelets
    b. pulses at all the major pulse points
    c. responsiveness
    d. any major bleeding and/or deformities

8. When applying a cervical collar, you should:
    a. Hold the head still with both hands while another rescuer applies the collar.
    b. Perform a head-tilt, chin-lift.
    c. Place one hand on the forehead of the patient while you apply the cervical collar with the other.
    d. Turn the patient's head from side to side to check for pain during movement.

9. The first goal of treatment for burn injuries is to stop the burning. To do that apply:
    a. copious amounts of ice over the burned area
    b. copious amounts of water over the burned area
    c. several ice packs
    d. short bursts from a CO_2 extinguisher directly to the burned area.

10. Using the *Rule of Nines,* what is the total percentage of body surface area burned on an adult male who has sustained a large burn on his entire left leg and left arm?
    a. 27%
    b. 19%
    c. 29%
    d. 18%

11. When calculating the percentage of burn surface on a 3-year-old, what is the percentage used for the head?
    a. 18%
    b. 16.5%
    c. 13.5%
    d. 9%

12. The most appropriate treatment for a fracture, dislocation, or sprain is which of the following?
    a. Only fractures get splinted.
    b. A dislocation cannot be splinted or immobilized because of the possibility of further injury.
    c. Regardless of the type of injury, the treatment is splinting and immobilization.
    d. Fractures and dislocations are more painful and should be splinted.

13. Prior to splinting an injury, check the extremity to be sure that blood is still flowing to the limb. The best method of doing this is to:
    a. Check for a pulse and/or capillary refill.
    b. Look at the skin color.
    c. Test the skin to see if it feels cooler.
    d. It is not important unless the bone has broken the skin.

14. To check capillary refill:
    a. Gently press down on the fingernail or toe until the skin turns white. After releasing the pressure, the nail bed should turn pink within 2 seconds.
    b. Gently compress the artery leading to the hand or foot until the nail beds become white. After releasing the artery, blood flow should make the nail beds pink in 2–3 seconds.
    c. Raise the injured extremity higher than the patient's heart. The nail beds should remain pink after 2 seconds.
    d. Check for a pulse, applying pressure for 2 seconds. If the nail beds turn white, release and count how long it takes for them to return to pink.

15. When faced with a medical emergency, your main focus should be to:
    a. Determine causes or diagnose medical conditions.
    b. Treat the symptoms displayed by the patient.
    c. Provide lift assists to advanced medical personnel.
    d. Keep open lines of communication between the different agencies working together on the patient.

16. SAMPLE stands for:
    a. signs and symptoms, allergies, medications, past medical history, last meal, events leading to the current problem
    b. seizures, agitation, medications, prior surgeries, last doctor appointment, examination findings
    c. symptoms, allergies, movement, pain, lost feeling/numbness, examination time
    d. scene safety, age, medications, past meal, loss of movement, extremity pains

17. Seizures are a common medical emergency. Patients who are actively seizing need advanced EMS care. Your initial care should include doing the following:
    a. Stay nearby so you are there to render aid when the patient stops seizing.
    b. Hold head stabilization to protect their neck from further injury.
    c. Place a bite stick in the patient's mouth to protect the tongue.
    d. Do not attempt to restrain the patient, never put anything in his or her mouth, protect their head from hitting hard surfaces, and move the patient to a safer location if possible.

18. Administering a patient's medication is permitted when:
    a. The patient is having a heart attack.
    b. The doctor has written it on the prescription.
    c. It is beyond your scope of training and therefore is not allowed.
    d. When the patient is unconscious.

19. What are the three main methods of controlling blood loss?
    a. direct pressure, elevation, and pressure points
    b. elevation, pressure points, and oxygen
    c. direct elevation, pressure points, and tourniquet
    d. direct pressure, oxygen, and pressure points

20. If bleeding persists and bandages become soaked with blood, you should:
    a. Remove all blood soaked dressings and bandages and apply new dressings to decrease risk of infection.
    b. Elevate the patient's feet and transport the patient to a medical facility.
    c. Place the patient on oxygen and wait for advanced care.
    d. Do not remove the blood-soaked dressings, but apply additional dressings on top.

21. Tourniquet application should be done using a commercial tourniquet device such as a _____.
    a. RIT
    b. CAT
    c. CAD
    d. T-shirt

22. While assisting a mother deliver her baby, you see the umbilical cord wrapped around the baby's neck. The most appropriate action would be to:
    a. Tell the mother to not push while you wait for advanced care to arrive.
    b. Do nothing, the cord stretches with the baby as it is born.
    c. Quickly clamp the cord and cut it to ensure the baby's safety.
    d. Gently unwrap the cord using two fingers to bring the cord forward and over the baby's upper body.

23. When examining a mother in labor you see a tiny foot emerging from the vaginal opening. How would you proceed?
    a. Do nothing, the mother needs to be transported immediately to the nearest medical facility.
    b. Continue to assist the mother on scene because the baby's birth is imminent.
    c. Gently push the foot back in, place her on her side and wait for advanced care.
    d. Pull gently, but firmly, on the baby's foot until the baby is born.

24. When it is time to cut the umbilical cord:
    a. Always use a commercial umbilical cord clamp.
    b. Wait for the arrival of providers with additional training
    c. Always cut the cord close to the baby.
    d. Take note of the exact time the cord was cut since that is the actual time of birth.

25. Your main objective as a firefighter on an EMS call is to:
    a. Make certain that all firefighters who are helping the patient are wearing PPE.
    b. Stabilize the patient prior to the arrival of personnel with a higher level of EMS training.
    c. Ensure that HIPAA is being followed.
    d. Ensure that record keeping of the event is being done so that proper documentation is kept.

CHAPTER 22 ANSWER KEY

Question #	Answer	Page #
1	b	692
2	d	694
3	a	695
4	c	695
5	a	696
6	a	696–697
7	d	696
8	a	697
9	b	698
10	a	698
11	a	698
12	c	689
13	a	689
14	a	699
15	b	699
16	a	697
17	d	700
18	c	700
19	a	704
20	d	704
21	b	704
22	d	706
23	a	706
24	b	706
25	b	707

Incident Command System

by Forest Reeder and Tim Flannery
with Alan Brunacini

1. Name two major objectives of an incident management system:

   1. hazard-zone management
   2. hazard-zone identification
   3. hazard-zone equipment tracking
   4. protection of hazard-zone workers

   a. 1 & 4
   b. 2 & 3
   c. 3 & 4
   d. 1 & 2

2. An incident that brings state-level personnel to the scene would be categorized by NIMS as what type of incident?

   a. Type 1
   b. Type 2
   c. Type 3
   d. Types 4 and 5

3. The type of response that accounts for more than 99% of all fire department incident activities is classified as what type of incident?

   a. Type 1
   b. Type 2
   c. Type 3
   d. Types 4 & 5

4. On the emergency scene, this person is the overall site manager, and everyone on the scene performs their assigned roles according to his or her plan:

   a. overhead manager
   b. NIMS coordinator
   c. incident commander
   d. scene boss

5. At this organizational level of scene management, one of the responsibilities of the incident commander is to develop and manage this kind of incident action plan:
   a. procedural
   b. task
   c. tactical
   d. strategic

6. At this level, the work on the fireground actually gets done:
   a. procedural
   b. task
   c. tactical
   d. strategic

7. An example of this organizational level in action would be a ladder company performing vertical ventilation on the roof so the engine company could safely operate on the interior:
   a. procedural
   b. task
   c. tactical
   d. strategic

8. Divisions represent _____, while groups represent _____.
   a. geographic locations, task assignments
   b. functional assignments, geographic locations
   c. tactical locations, functional assignments
   d. geographic locations, functional assignments

9. This person is responsible for assuming the role of the first incident commander:
   a. first subordinate chief officer on scene
   b. first unit or member arriving at the incident scene
   c. second unit or member, if the first unit or member goes into the fast attack mode
   d. first superior chief officer on the scene

10. Before employing a strategy, the incident commander must take what action?
    a. Conduct a size-up of the critical incident factors.
    b. Assign the first-arriving units.
    c. Secure the most effective water supplies.
    d. Create an incident action plan.

11. How does the incident commander begin the communication process?
    a. Assign a radio frequency to the incident.
    b. Initiate a transmission of orders for the first-arrivers.
    c. Transmit a standard initial radio report.
    d. Control the transmission process.

12. This is one of the cornerstones of IMS and allows companies to wait in an uncommitted manner until assigned by the incident commander:
    a. response protocol
    b. staging procedures
    c. scene stalling
    d. prearrival procedures

13. The primary focus of defensive fires is cutting off the main body of fire and
    a. ensuring collapse zones are maintained
    b. surrounding and drowning
    c. striking additional alarms
    d. protecting exposures

14. Which method of transfer of command allows the most complete incident communication?
    a. by telephone
    b. face-to-face
    b. over the radio
    d. by text

15. On the command team, this person manages the tactical worksheet and verifies that the correct strategy and IAPs are being used:
    a. command technician
    b. senior advisor
    c. support officer
    d. tracking officer

16. This ICS position manages all resources that are required for support of the incident:
    a. safety
    b. logistics
    c. planning
    d. administration

17. This is the IC's primary responsibility on any incident:
    a. Document the activities of the incident.
    b. Ensure the safety of all members working the incident.
    c. Mitigate the incident in the quickest manner possible.
    d. Manage the incident and set the action plan.

18. When does the incident commander submit an "under control" report?
    a. when the forward progress of the fire is stopped
    b. when smoke conditions have been eliminated
    c. when the fire is definitely out
    d. when salvage operations have been completed

19. Damage caused by firefighting operations such as breaking windows and cutting roofs is known as:
    a. indirect damage
    b. collateral damage
    c. primary damage
    d. secondary damage

20. This type command is typically used where multiagency or multijurisdictional cooperation is required:
    a. dual-phase command
    b. unified command
    c. strike force command
    d. area command

21. This officer of the general staff provides accurate and complete information about the incident to the media and the public:
    a. liaison officer
    b. logistics section chief
    c. public Information officer
    d. intelligence officer

22. This incident resource is composed of resources of the same kind and type:
    a. single resource
    b. strike team
    c. task force
    d. division

23. This incident resource is a mix of different kinds of resources assembled for a particular tactical need:
    a. group
    b. strike team
    c. task force
    d. division

24. This method of organizing tactical operations is to assign resources geographically:
    a. group
    b. branch
    c. task force
    d. division

25. This method of organizing tactical operations is accomplished by assigning personnel to specific purposeful activities:
    a. group
    b. branch
    c. task force
    d. division

26. When the incident commander finds that the span of control for the number of divisions and/or groups in place has become unwieldy or unmanageable, what should be done?
    a. A logistics section chief should be assigned.
    b. Branches should be created.
    c. The operations section chief should directly oversee the operation.
    d. An incident command post should be established.

27. This ICS facility is where the incident commander oversees the incident:
    a. base camp
    b. main base
    c. command post
    d. command base

28. In this type of staging, the incident commander determines which units are needed on scene and instructs the rest to stop approximately one block from the scene:
    a. auxiliary staging
    b. Level 1 staging
    c. Level 2 staging
    d. directed Stand-by

29. At this incident facility, the functions of the logistics section and the finance/administration sections are coordinated:

    a. unified command post

    b. base

    c. camp

    d. command post

30. This incident facility is where resources may be kept to support incident operations when very large areas of operation are involved:

    a. unified command post

    b. base

    c. camp

    d. command post

CHAPTER 23 ANSWER KEY

Question #	Answer	Page #
1	a	715
2	c	716
3	d	716
4	c	716
5	d	717
6	b	717
7	c	717
8	d	719
9	b	719
10	a	727
11	c	726–727
12	b	727
13	d	726
14	b	728
15	c	728
16	b	724
17	b	725
18	a	726
19	d	726
20	b	714
21	c	718
22	b	719
23	c	721
24	d	719
25	a	719
26	b	720
27	c	723
28	c	724
29	b	724
30	c	724

Advanced Communications

by Charles Jennings

24

Advanced Communications

Questions 1–4 are based on the communication model developed by Shannon and Weaver.

1. A message is sent via a radio frequency, telephone line, and the like, which is known as a(n) _____
   a. transmitter
   b. channel
   c. receiver
   d. information source

2. This point of the communications model encodes the message:
   a. transmitter
   b. channel
   c. receiver
   d. information source

3. When the message reaches this point, it is decoded:
   a. transmitter
   b. channel
   c. receiver
   d. information source

4. A message is produced by which of the following?
   a. transmitter
   b. channel
   c. receiver
   d. information source

5. You are transmitting over the radio in close proximity to other personnel. Feedback is being produced. What can you do to rectify this problem?
   a. Ask them to turn their radios off.
   b. Ask them to switch to another frequency.
   c. Ask them to cover their microphones.
   d. Ask them to move to another area.

6. Before transmitting, why should radio users listen carefully?
   a. to ensure that they are not being called by command with an assignment
   b. to ensure that they are on the proper frequency
   c. to ensure that they are not interrupting between two other parties
   d. to ensure that their transmission will not be interrupted by feedback

7. Suppose you are a firefighter working apart from your crew in a high-rise fire. They are on the landing above, preparing the attack line. You are assigned the standpipe control valve position on the floor below the fire. You are equipped with a portable radio. A civilian tells you that there is an elderly man on the fire floor who uses a walker. He tells you that the elderly man has probably not been evacuated. What action do you take?
   a. Go up to the landing above and tell your supervisor face-to-face.
   b. Notify dispatch of the situation and the need for assistance.
   c. Use your portable radio to notify the highest ranking person on your channel.
   d. Explain to the civilian that in a high-rise fire the elderly man is best off staying where he is.

8. The ability of emergency responders to work seamlessly with other systems or products without any special effort is the definition of:
   a. multiplicity
   b. organizational balance
   c. interoperability
   d. interagency coordination

9. These are devices that are used to link different frequencies together at a specific location and are often mobile and carried in a command or communications vehicle:
   a. repeaters
   b. gateways
   c. network coordinators
   d. automated dispatchers

10. Regarding mobile communications vehicles, which of the following is false?
    a. They are excellent places for the incident commander to operate.
    b. They may be as simple as a work surface and radio with mutual aid frequencies incorporated into a chief's vehicle.
    c. These vehicles should be equipped with generators and portable lighting, and be capable of operating for extended periods.
    d. Units may now include a link to the computer-aided dispatch system, mapping software, and capabilities for receiving television and radio broadcasts.

11. Although some jurisdictions elect not to participate, utilizing this reporting system allows departments to have consistent data for comparison with other organizations and permits analysis of response times, staffing, and fire losses in one database:

    a. National Fire Database (NFD)

    b. American Incident Analysis Forum (AIAF)

    c. National Fire Incident Reporting System (NFIRS)

    d. National Emergency Response Database (NERD)

12. Regarding the reporting system referred to in question 27, what information provided by that system would be most important to firefighters?

    a. staffing requirements

    b. trends in firefighter casualties

    c. resource allocation

    d. apparatus location and response times

13. One of the more extensive reports prepared by the fire department itself is completed after a major incident that resulted in death(s) or major property damage and usually contain detailed descriptions of what took place. What is this report called?

    a. incident supplement report

    b. after action report

    c. incident investigation report

    d. post-incident analysis report

14. This type of reporting system is defined as the combined use of information technology and electronics (usually in mobile settings) and will be used to generate reports of lost, stolen, or damaged vehicles directly to 9-1-1 centers:

    a. telematics

    b. autotronics

    c. satellite imaging

    d. radiographics

CHAPTER 24 ANSWER KEY

Question #	Answer	Page #
1	b	735–736
2	a	735–736
3	c	735–736
4	d	735
5	c	740
6	c	740
7	c	738
8	c	742
9	b	742
10	a	742
11	c	736–738
12	b	738
13	b	738
14	a	739

Pre-Incident Planning

by Jack J. Murphy

25

1. A pre-incident plan gives the fireground commander:
   a. something to refer to when needed
   b. an unchangeable standard that must be followed
   c. a workable plan for all contingencies
   d. inside information about the building contents

2. To improve their tactical capabilities, firefighters should improve their:
   a. testing skills
   b. building intelligence gathering skills
   c. communication skills
   d. driving skills

3. Ignition, burning, and fire spread within a building relates to the characteristics of:
   a. combustible materials present in the building
   b. the building's structural elements
   c. the building's designed layout
   d. the building's fire protection system, if present

4. The best way to address a fire is by performing an exterior reconnaissance to note:
   a. nearby water supply sources
   b. forcible-entry problems
   c. access roadways
   d. exposure issues

5. When formulating plans on water flow, internal building reconnaissance should include:
   a. hydrant locations
   b. detection and audible alarm systems
   c. standpipes and sprinkler systems
   d. information on the supplied domestic water

6. Pre-incident planning helps in performing primary searches because:
   a. Occupant location will be known.
   b. The floor design will be known.
   c. The incident commander can refer to the preplan while the firefighters are searching.
   d. Company officers will be familiar with the preplan.

7. Multilock doors, metal gates, and new locking devices should be noted on reconnaissance plans because:
   a. Entrance locations should be known beforehand.
   b. Additional units may need to be called.
   c. They offer more resistance to forcible entry.
   d. It may require a change in hoseline selection.

8. Regarding hand line selection, pre-incident planning helps with:
   a. hoseline stretching
   b. hoseline placement within a building
   c. sufficient water discharge
   d. all of the above

9. Regarding ventilation, pre-incident planning should:
   a. Identify paths of smoke travel.
   b. Give the chief a good idea of how many ladder companies are required.
   c. Give the company officer the appropriate tool selection.
   d. Describe how the ventilation must be carried out.

10. *Target hazards* is a term used to describe:
    a. buildings that present a significant life safety concern
    b. buildings that present a significant firefighting challenge
    c. buildings that have been subject to numerous false alarms
    d. both A and B

11. Gathering building intelligence prior to an incident greatly helps in reducing:
    a. building costs
    b. firefighter deaths
    c. the number of fires
    d. manpower

12. The best way for responding firefighters to access gathered building information is through the use of department:
    a. binders kept in chiefs' cars
    b. paper plans available in the apparatus
    c. mobile computer terminals
    d. firehouse-based computer systems

13. Exhibition halls, theaters, and stadiums are examples of which type of use group?
    a. business
    b. assembly
    c. mercantile
    d. high hazard

14. An occupancy that processes materials that constitute a health hazard exceeding limits allowed by building and fire codes is said to be a:
    a. factory use group
    b. high hazard use group
    c. institution use group
    d. mercantile use group

15. Day care centers, nursing homes, and assisted living facilities are examples of which type of use group?
    a. institution
    b. high hazard
    c. business
    d. assembly

16. Which of the following is an example of a low hazard in a storage facility use group?
    a. aircraft hanger
    b. aerosol storage
    c. book storage
    d. fur storage

17. When conducting building reconnaissance, an exterior building survey should include all of the following except:
    a. number of stories
    b. type of occupancy
    c. occupant load
    d. key box location

18. For the site survey, interior floors should be identified as:
    a. levels
    b. sectors
    c. floors
    d. divisions

19. Balloon and platform are examples of which type of construction?
    a. ordinary
    b. wood frame
    c. noncombustible
    d. heavy timber

20. A stairway that is open between two floors in a common occupancy, and separate from the main stairwell serving the building, is known as a(n):
    a. access staircase
    b. fire staircase
    c. enclosed staircase
    d. common staircase

21. Newly constructed buildings may include areas of refuge for disabled people, called:
    a. safe refuge rooms
    b. designated rescue locations
    c. designated assistance points
    d. areas of rescue assistance

22. When preplanning water supply available from private hydrants and water tanks, which of the following should be identified:
    a. tank capacity, drafting locations, and who is responsible for maintenance
    b. size of hydrant outlets, size of water mains, and tank capacity
    c. size of the water main, tank capacity, and who is responsible for system maintenance
    d. tank capacity, location of water mains, and size of hydrant outlets

23. Multiplying a buildings length by its width and dividing that number by 3 will provide the answer for:
    a. total gallons needed per minute to extinguish one floor of fire
    b. total gallons needed per minute to extinguish one room of fire
    c. total gallons needed per minute to extinguish every two floors of fire
    d. total gallons needed per hour to extinguish one floor of fire

24. To figure out water supply needs when considering an exposure building on fire, add _____ of the 100% building involvement figure for the original gpm required.
    a. 25%
    b. 50%
    c. 75%
    d. 100%

25. Dry and preaction describe two types of:
    a. building water main supply systems
    b. building standpipe systems
    c. building alarm systems
    d. building sprinkler systems

26. The role of a building's fire pump is to:
    a. Pressurize a sprinkler or standpipe system with air to remove debris.
    b. Compensate for any pressure reducing valves in the system.
    c. Increase water pressure for a building's standpipe and sprinkler system.
    d. Remove water delivered from fire hoses that have flooded the basement.

27. The fire department can augment a standpipe system by supplying more water through the building's _____ if available.
    a. dry hydrant
    b. fire department connection
    c. fire pump
    d. domestic water connection

28. Computer data centers, dip tanks, and electronic equipment areas may be protected by:
    a. pressurized water extinguishing systems
    b. chemical extinguishing systems
    c. halon extinguishing systems only
    d. fixed oxygen depletion units

29. The difference between a firewall and a fire partition is that:
    a. A firewall separates one building into two, and a fire partition restricts fire spread from one area of the building to another.
    b. A fire partition separates one building into two, and a firewall restricts fire spread from one area of the building to another.
    c. Only openings in firewalls need to be protected.
    d. Only openings in fire partitions need to be protected.

30. When preplanning for hazardous materials risks within a building, hazardous material documentation, including _____, should be reviewed.
    a. Hazardous Material Reporting Logs
    b. Hazardous Material Data Logs
    c. Material Safety Data Sheets
    d. Material Hazards Data Sheets

31. A building footprint map provides building intelligence to this interior location:
    a. roof style
    b. fire escape
    c. fire hydrant
    d. stairway with standpipe hose connection

32. When collecting building data, the building's structural members should be identified according to the type of material supporting the:
    a. horizontal load of the roof and floor
    b. vertical load of the roof and floor
    c. exterior walls only
    d. floors only

33. If the risk is too great to perform building reconnaissance on the interior of a vacant structure, firefighters should conduct an exterior ground level recon as well as a(n):
    a. exterior roof recon
    b. tower ladder observation
    c. interview with the building owner
    d. truck observation

34. The most important reason information should be gathered on existing, occupied structures undergoing renovations is because:
    a. Exit paths may be altered, increasing the threat to life safety.
    b. Utilities may be shut down to certain areas.
    c. Office partition walls may be changed.
    d. Workers in the building may be unfamiliar with the building's total layout.

35. When gathering intelligence about standpipe systems for buildings under construction, firefighters should note the accessibility of the fire department connection, that the standpipe is within two floors of the highest level of work, and that:

    a. The top of the riser is not left open without a shutoff valve.

    b. The floor outlets valves are closed, especially at the below-grade level.

    c. The fire pump can be operated.

    d. both a and b

36. Heating and cooling building management systems can assist the incident commander with regard to smoke control within a structure by:

    a. providing cool, smoke free areas for firefighter rehabilitation

    b. allowing the affected area to be shut down from a remote location

    c. only requiring minimal personnel into the fire area for controlling smoke flow

    d. removing the need for horizontal window ventilation

37. If a building is equipped with elevators, the preplan should indicate the location of the level where the firefighter phase II key access control is located, as well as the location of the:

    a. elevator mechanical room and the pit room

    b. elevator mechanical room only

    c. elevator control room

    d. pit room access

38. A preplan should include information about the existence of a cockloft. A cockloft is best described as the:

    a. void space running horizontally between floor beams

    b. void space running vertically between inner finished walls and exterior walls

    c. void space between the roof and the top floor ceiling

    d. void space between the underside of lowest floor and the ground

39. When conducting water supply fire flow testing, all of the following are required, except:

    a. two fire hydrants

    b. a map of water supply piping to the hydrants

    c. a ruler

    d. a fire pumper truck

CHAPTER 25 ANSWER KEY

Question #	Answer	Page #	Question #	Answer	Page #
1	d	747	31	d	758
2	b	747–748	32	a	759
3	a	748	33	b	763
4	a	748	34	a	763
5	c	748	35	d	763
6	b	748	36	b	773
7	c	748	37	a	759
8	d	748	38	c	759
9	a	748	39	d	756
10	d	749			
11	b	749			
12	c	749			
13	b	749			
14	b	750			
15	a	750			
16	a	750			
17	c	751			
18	d	751			
19	b	752			
20	a	752			
21	d	752			
22	c	753			
23	a	753			
24	b	753			
25	d	757			
26	c	757			
27	b	757			
28	b	757–758			
29	a	758			
30	c	758			

Fire Protection Systems

by Bob Till

1. Buildings protected by fire alarm systems are called:
   a. proprietary premises
   b. protected premises
   c. insured premises
   d. guarded premises

2. This type of alarm signal provides a way of monitoring an alarm system status and indicating abnormal conditions:
   a. remotely transmitted signal
   b. trouble signal
   c. local alarm signal
   d. supervisory signal

3. This type of alarm signal notifies the fire department directly or indirectly of fire:
   a. remotely transmitted signal
   b. trouble signal
   c. local alarm signal
   d. supervisory signal

4. This type of alarm signal notifies a building's occupants of fire:
   a. remotely transmitted signal
   b. trouble signal
   c. local alarm signal
   d. supervisory signal

5. This type of alarm signal identifies system problems that can influence the system's reliability:
   a. remotely transmitted signal
   b. trouble signal
   c. local alarm signal
   d. supervisory signal

6. This key component of a fire alarm system is an automatic or manual mechanism that causes a system to indicate an alarm condition:
    a. alarm notification appliance
    b. alarm-initiating device
    c. fire alarm control panel
    d. alarm annunciator

7. This key component of a fire alarm system is a mechanism that makes building occupants aware of an alarm condition:
    a. alarm notification appliance
    b. alarm-initiating device
    c. fire alarm control panel
    d. alarm annunciator

8. The method used to connect and control alarm-initiating devices helps to determine:
    a. how quickly the alarm is sent to the fire dispatcher
    b. what auxiliary systems for powering the equipment must be in place
    c. what a fire department knows about the fire when firefighters arrive
    d. how often the system must be serviced

9. These alarm-initiating devices use similar ways to detect fire at a particular point or location. Typically, these devices are heat, smoke, or rate-of-rise detectors:
    a. manual pull stations
    b. linear detectors
    c. smoke sensors
    d. spot detectors

10. What is the most common form of heat detector?
    a. sprinkler head
    b. rate-of-rise detector
    c. smoke detector
    d. fixed-temperature detector

11. This type heat detector activates when the temperature surrounding it rises faster than a predetermined value:
    a. rate-of-rise detector
    b. rate compensation detector
    c. rate differential detector
    d. heat compensation detector

12. This kind of heat detector can be particularly useful in spaces where large temperature changes or cooling air movements occur:

    a. rate-of-rise detector

    b. rate compensation detector

    c. rate differential detector

    d. heat compensation detector

13. The type of smoke alarm is typically used in existing private homes without detection-system code requirements. These units run on batteries and have built-in alarms. When they are activated, a local alarm sounds.

    a. combination smoke alarm device

    b. single-station smoke alarm device

    c. linear-pole smoke alarm device

    d. multiple-link smoke alarm device

14. Regarding smoke alarm requirements for new residences, which of the following statements for device locations is not true?

    a. At least one device must be located on each level.

    b. At least one device must be located in each bedroom.

    c. One device must be located in the corridor outside each sleeping area.

    d. Corridor devices must be interconnected so that all corridor devices will sound when one is activated.

15. This type of photoelectric smoke detector is used to cover large areas such as atriums and warehouses:

    a. ionization

    b. light scattering

    c. light obscuration

    d. projected beam

16. This type of smoke detector is mounted inside electrical cabinets and other spaces where rapid detection is necessary to prevent damage:

    a. projected beam

    b. air sampling

    c. gas detector

    d. ionization

17. Welding or sunlight can cause unwanted activation of this type of detector:
    a. spot detector
    b. gas detector
    c. flame detector
    d. air-sampling detector

18. Alarm verification systems can prevent frequent false activation of systems, which can result in occupants ignoring alarms during an actual fire. These systems are typically found in what kind of structure?
    a. residential
    b. commercial
    c. educational
    d. industrial

19. An alarm system in which two detection devices must activate before initiating a fire alarm condition is called a _____.
    a. redundant alarm system
    b. dual-zone alarm system
    c. cross-zoned alarm system
    d. dual-detection alarm system

20. This type of alarm system is usually created by connecting the set of fire detection devices inside the protected building to a master box located on the outside of the building:
    a. auxiliary fire alarm system
    b. local fire alarm system
    c. central station alarm system
    d. proprietary alarm system

21. This type of alarm system sounds an alarm within the building where it is activated but does not send a signal out of the building:
    a. auxiliary fire alarm system
    b. local fire alarm system
    c. central station alarm system
    d. remote supervising-station alarm system

22. This type of alarm system transmits a signal from a protected facility to a monitoring station operated by the facility's owner. These systems are common on college and office campuses, where the same group manages and is responsible for protecting many buildings:

    a. remote supervising-station alarm system

    b. local fire alarm system

    c. central station alarm system

    d. proprietary alarm system

23. This type of alarm system utilizes a location that is staffed by operators who receive a signal and take action as required by the National Fire Alarm Code, such as notifying the fire department:

    a. auxiliary fire alarm system

    b. local fire alarm system

    c. central station alarm system

    d. proprietary alarm system

24. This type of alarm system transmits supervising and trouble signals to a maintenance facility, while any alarm signals go to an alarm-dispatch facility:

    a. auxiliary fire alarm system

    b. remote supervising-station alarm system

    c. central station alarm system

    d. proprietary alarm system

25. In this type of alarm coding category, the fire alarm control panel shows the zone where the activated device is located, and it may indicate the specific device that was activated:

    a. noncoded alarm

    b. zoned noncoded alarm

    c. zoned coded alarm

    d. master coded alarm

26. These systems can extinguish or control fires without firefighter intervention and therefore save lives and property:

    a. fire detection

    b. some control

    c. fire alarm

    d. fire suppression

27. Regarding working with fire alarm systems, which of the following is not true?
    a. Do not reset the system until the device of origin has been determined.
    b. Use the silence switch on the panel only if you are sure that the activation is a false alarm.
    c. When searching for the device of origin, look for the detector that is showing a blinking red light on the individual detection device.
    d. A system with addressable devices will tell you the exact device that has been activated at the fire alarm control panel.

28. How do sprinkler systems improve life safety?
    a. by ensuring prompt notification of fire-suppression forces when activated
    b. by reducing smoke production and spread in the fire area
    c. by preventing flashover and fire spread outside the room of origin
    d. by preventing the ignition of fires in residential structures

29. On a sprinkler head, this device breaks the water stream into droplets and pushes them to cover an area larger than a straight stream of water would:
    a. impeller
    b. deflector
    c. nozzle
    d. grooved cap

30. Which of the following is not a type of sprinkler head?
    a. upright
    b. gravity
    c. sidewall
    d. pendant

31. This type of sprinkler head has no release mechanism and is sometimes used in special situations, for example, to apply foam to a hazard:
    a. residential sprinkler head
    b. dry sprinkler head
    c. early-suppression, fast-response sprinkler head
    d. deluge sprinkler head

32. In this type of sprinkler system, which is commonly used when freezing is a consideration, the piping is filled with air or nitrogen instead of water:
    a. wet-pipe system
    b. dry-pipe system
    c. preaction system
    d. deluge system

33. This type of system minimizes water damage and requires an additional detection system and a sprinkler head's fusing for water to flow:
    a. wet-pipe system
    b. dry-pipe system
    c. preaction system
    d. deluge system

34. This is a system of pipes that flow water onto a fire when an automatic sprinkler head is fused:
    a. wet-pipe system
    b. dry-pipe system
    c. preaction system
    d. deluge system

35. This system consists of pipes with open sprinkler heads where all heads flow when the system is activated:
    a. wet-pipe system
    b. dry-pipe system
    c. preaction system
    d. deluge system

36. This type of valve permits water flow in only one direction:
    a. singular valve
    b. check valve
    c. full flow valve
    d. monitor valve

37. In a _____ system, the _____ valve opens only when the air pressure drops.
    a. wet-pipe, check
    b. dry-pipe, check
    c. dry-pipe, dry-pipe
    d. deluge, dry-pipe

38. In a deluge sprinkler system, the type of initiating device is based on the:
    a. hazard being protected
    b. area being protected
    c. proximity to the water supply
    d. type of heads on the system

39. Which of the following is not a main objective of a residential sprinkler system?
    a. preventing flashover in rooms of origin
    b. preventing structural collapse
    c. improving chances for occupant escape
    d. preventing fire spread to adjacent rooms

40. This smaller capacity high-pressure pump is sometimes installed next to the main fire pump. Its job is to maintain the pressure upstream of the pump so that small pressure variations do not cause main fire pump activation.
    a. auxiliary pump
    b. secondary pump
    c. supplemental pump
    d. jockey pump

41. What is the major reason for sprinkler system failure?
    a. The main water supply control valve has been left shut accidentally.
    b. The fire department failed to supplement the system.
    c. The water supply serving the system is inadequate.
    d. The piping in the system has been designed improperly.

42. This type of water supply control valve exposes a stem to indicate that it is open:
    a. post-indicator valve
    b. OS&Y valve
    c. wall post-indicator valve
    d. wall valve

43. During sprinkler operations, when should the riser control valve be closed?
    a. when the fire is under control
    b. when there are firefighters with portable radios available to monitor the valve
    c. upon orders of the incident commander
    d. when the fire is extinguished

44. This type of standpipe is designed for use by building occupants. They typically do not have a fire department connection (FDC), and they operate at pressures and flows that are usually too low to be of use to firefighters.
    a. Class I
    b. Class II
    c. Class III
    d. Class IV

45. This type of standpipe has outlets for fire department use and a preconnected hose for building occupants:
    a. Class I
    b. Class II
    c. Class III
    d. Class IV

46. This type of standpipe outlet has a 2.5-in. male coupling and a valve to control water flow. Hose outlet valves connect to attack lines, and personnel within a burning structure can deliver water.
    a. Class I
    b. Class II
    c. Class III
    d. Class IV

47. What type of pressure-control valve commonly utilizes an orifice plate?
    a. pressure restricting
    b. pressure reducing
    c. pressure control
    d. pressure monitoring

48. You are supplying an FDC for a standpipe. You find that the connection is clogged. What options might you consider in this situation?
    a. Supply the sprinkler instead.
    b. Hand stretch the hoselines into the building and up the stairs.
    c. Connect to and pump through a hose valve in a stairwell.
    d. Use a hand tool to unclog the pipe.

49. When exposed to flame, halocarbons usually produce this gas, which is not safe for humans to breathe:
    a. hydrochloric acid
    b. bromine gas
    c. hydrogen tetrachloride
    d. hydrofluoric acid

50. This type of extinguishing system is often used in restaurant hood systems to extinguish fires in deep fat fryers and other cooking equipment:
    a. dry chemical
    b. wet chemical
    c. carbon dioxide
    d. fluorine

CHAPTER 26 ANSWER KEY

Question #	Answer	Page #	Question #	Answer	Page #
1	b	768	31	d	779
2	d	768	32	b	779
3	a	768	33	c	779
4	c	768	34	a	779
5	b	768	35	d	779
6	b	768	36	b	778
7	b	768	37	c	778
8	c	768	38	a	779
9	d	768	39	b	780
10	a	769	40	d	780–781
11	a	769	41	a	782
12	b	770	42	b	781
13	b	770	43	c	782
14	d	770	44	b	783
15	d	771	45	c	784
16	b	771	46	a	783
17	c	772	47	a	783
18	b	772	48	c	784
19	c	772	49	d	786
20	a	773	50	b	768–787
21	b	773			
22	d	774			
23	c	774			
24	b	774			
25	b	775			
26	d	775			
27	c	775			
28	c	776			
29	b	776			
30	b	777–778			

Advanced Fire Attack

by Christopher Flatley with Jerry Knapp

27

1. As you become a more seasoned firefighter, you will be tasked with greater responsibility. What comes with that responsibility?
   a. danger
   b. accountability
   c. credibility
   d. judgment

2. This enables firefighters to make the best possible decision on how to deal with a situation:
   a. preaction plans
   b. SOPs
   c. personal protective equipment
   d. size-up

3. You suspect that there is a fire in the basement of a building. What action can be taken to confirm this suspicion?
   a. Break out small pane of glass in a basement window.
   b. Feel the first floor for evidence of heat.
   c. Open the basement door from the interior and descend.
   d. Check the upper floors for fire extension.

4. Regarding life hazard, *occupied* refers to a home that someone lives in. If no one was home, the house would be referred to as being _____.
   a. vacant
   b. derelict
   c. unoccupied
   d. undetermined

5. When sizing up a building from the outside, you notice windows that are off height from others on the same side of the house. What would this indicate to you?
   a. a bedroom
   b. a bathroom
   c. a kitchen
   d. a stair half-landing

6. The goals for stretching hoselines are as follows (put in correct order):

   | 1. to protect property |
   | 2. to confine the fire |
   | 3. to protect life |
   | 4. to extinguish the fire |

   a. 1, 2, 3, 4
   b. 2, 3, 1, 4
   c. 3, 1, 2, 4
   d. 3, 2, 1, 4

7. The most important considerations when stretching a hoseline are to make sure to select the correct size hoseline and to make sure that:
   a. It is properly flaked out.
   b. It reaches the fire before the line is charged.
   c. It is properly pressurized with water.
   d. It is backed up by another hoseline.

8. Why does the wood frame construction typically used in garden apartments play a big role in fire attack?
   a. Large void spaces run the length of the structure.
   b. Wood staircases, usually on the exterior, are prone to vertical fire spread.
   c. Combustible exterior walls are vulnerable to lateral fire spread.
   d. Roofing materials are conducive to ignition by flying brands.

9. When forcing entry into apartments in a garden apartment complex, which is the most correct statement?
   a. Open any doors that may be necessary.
   b. Open only the fire apartment door.
   c. Open all doors in the complex.
   d. Open only doors on the fire floor.

10. In a tenement with a railroad flat layout, where would you find the secondary means of egress?
    a. secondary stairs at the rear
    b. front fire escape
    c. rear fire escape
    d. front and rear fire escapes

11. In an H-type multiple dwelling, this is the best type stair layout from a fire department's point of view:
    a. isolated stairs
    b. wing stairs
    c. transverse stairs
    d. semi-isolated stairs

12. Successful firefighting in multistory multiple dwellings is often based on the results of:
    a. completion of a trench cut between wings
    b. choosing the proper attack route
    c. proper ventilation
    d. aggressive search operations

13. In large multiple dwellings, this type ventilation must be coordinated with hoseline advance:
    a. positive pressure ventilation
    b. vertical ventilation
    c. horizontal ventilation
    d. fog ventilation

14. Regarding trench cuts, which of the following is not correct?
    a. The trench cut becomes a fire break between fire walls.
    b. The trench should not be pulled until fire has entered the cockloft.
    c. Hoselines can be operated into the trench.
    d. The trench can be used as a second vent hole.

15. Which tools should firefighters use when entering commercial buildings?

   1. thermal imaging camera
   2. a search line
   3. a set of forcible-entry irons
   4. a hoseline

   a. 1 & 2
   b. 2 & 4
   c. 1 & 4
   d. 3 & 4

16. In commercial structures, these walls depend on the roof system to hold them upright:
    a. panel walls
    b. block walls
    c. veneer walls
    d. tilt slab walls

17. Why is cutting a corrugated metal roof deck not recommended?
    a. The deck can melt underneath the firefighters.
    b. A section can hinge downward and drop a firefighter.
    c. The covering on the decking is corrosive.
    d. Sparks from the saw blade can ignite the combustible deck covering.

18. You are cutting a flat roof on a commercial building and as you sink in the blade, a white powdery residue issues from the cut. What do you do?
    a. Put the saw aside. Use a sledgehammer to break open the roof.
    b. Continue to cut, but do not breathe the residue.
    c. Stop cutting, notify command, and prepare to evacuate.
    d. Stop cutting, notify command, and request a hoseline to cut down on the dust.

19. What is the best way to ventilate a commercial building?
    a. Open natural ventilation points on the roof.
    b. Remove available windows.
    c. Do not vent until the fire is under control.
    d. Utilize positive pressure ventilation exclusively.

20. A floor layout where all mechanical equipment, electrical equipment, and stairwells are located in the center of the floor plan is known as:

    a. perimeter design

    b. perimeter core

    c. central core

    d. center core

21. You have responded to a high-rise office building and have a reported smoke condition on the 21st floor. You are the elevator operator. What floor would you take the elevator to?

    a. 18th floor

    b. 19th floor

    c. 20th floor

    d. 21st floor

22. Before any glass is broken in a high-rise, what action should be taken?

    a. The incident commander should be notified so the street can be secured.

    b. Pressure devices should be removed to alleviate window pressure.

    c. Large-diameter hoselines should be actively flowing water on the fire.

    d. An adhesive should be put on the glass so it can be removed to the interior.

23. You are on an engine company and have responded to a fire in a high-rise on floor 17. Regarding fire attack, which is the best answer?

    a. Hook up the initial attack line to a hose cabinet on floor 17.

    b. Hook up to a standpipe connection in the stairwell on floor 17.

    c. Hook up to a standpipe connection in the stairwell on floor 16.

    d. Hook up to a standpipe connection in the stairwell on floor 15.

24. You have responded to a high-rise fire. You are at the hallway door on the fire floor. There are occupants coming down the stairs from above and firefighters making their way up to the upper floors. What action do you take in regard to attack?

    a. Delay the attack until the stairwell in clear.

    b. Open the door and advance the attack line to the fire apartment.

    c. Open the door and use a wide fog stream to push the products of combustion away from the stairwell.

    d. Switch the attack to a different stairwell.

25. As an engine company firefighter in a commercial kitchen fire in a restaurant, what action should be taken before operating a hoseline?
    a. Ensure that foam is available as a backup.
    b. Ensure that the nozzle pattern is set on wide fog.
    c. Ensure that the utilities are controlled.
    d. Ensure that there is a vent path opposite the hoseline.

26. You have operated a hoseline in a kitchen of a restaurant and the floor has become extremely slippery. What action can be taken to improve traction?
    a. Use more water to further dilute any grease on the floor.
    b. Dump out salt shakers or bulk salt containers on the floor.
    c. Put down newspapers or linens to soak up any accumulated water or grease.
    d. Take off your boots and use your work shoes.

27. Operations at fires in vacant buildings must be conducted methodically with the emphasis on:
    a. determining the cause
    b. exposure protection
    c. firefighter safety
    d. void awareness

28. When attempting to control a natural gas leak, the first place to attempt that control is:
    a. at the affected appliance
    b. at the valve on the main line where it enters the building
    c. at the curb box in the street
    d. at the gas meter in the cellar or basement

29. To shut the gas off in a structure, how do you operate handle of the shutoff valve?
    a. Turn it until it is inline with the pipe it is on.
    b. Turn the valve two full turns clockwise.
    c. Turn the valve until it is perpendicular with the pipe it is on.
    d. Turn the valve one full turn counterclockwise.

30. You are attempting to operate a valve to turn off a gas service to a structure. It has been exposed to the elements and will not turn with a wrench. What do you do?
    a. Use a larger pipe wrench with a cheater bar.
    b. Do not force. Summon the utility company.
    c. Spray a CO_2 extinguisher on the valve. Wait 1 minute. Try to loosen it.
    d. Force the valve. Its breakaway feature will trip the valve to close.

31. Once a gas leak in building has been controlled, the next action should be to:
    a. Keep the building sealed for 1 hour to check for gas accumulation.
    b. Ventilate, starting at the lowest point in the building.
    c. Use vent fans to clear the structure.
    d. Ventilate, starting at the highest point in the building.

32. You respond to an incident involving a vehicle powered by compressed natural gas (CNG). You have been ordered to stop the flow of CNG. Where would you look to find the main shutoff mechanism and carry out this order?
    a. under the rocker panel
    b. near the fuel cylinder
    c. under the dashboard
    d. at the ignition, turn off the vehicle

33. Suppose you respond to a CNG-powered vehicle that is on fire. What action would you take?
    a. Shut off the gas at the control valve.
    b. Find and remove the CNG cylinder.
    c. Let the fire burn and protect exposures.
    d. Attack the fire from the flanks.

34. What would be the first action to take when confronted with an inside propane leak with no fire?
    a. Find the leaking cylinder.
    b. Eliminate sources of ignition.
    c. Ventilate the house.
    d. Close the valve.

35. With all cylinders involved in fire, the best course of action is to:
    1. Apply water large amounts of water from a safe distance.
    2. Move in under wide fog stream protection and shut the valve.
    3. Move the tank to a safe area.
    4. Allow the contents to burn off while protecting exposures.

    a. 1 & 4
    b. 2 & 3
    c. 1 & 2
    d. 3 & 4

36. The extinguishing agent of choice for engine companies when confronted with flammable liquid spills and fire is:
    a. carbon dioxide (CO_2)
    b. Class B foam
    c. dry chemical
    d. water

37. For a three-dimensional fire, what extinguishing agent should be used?
    a. foam first, then dry chemical
    b. foam only
    c. dry chemical first, then foam
    d. foam first, then water

38. Water should not be used on chemicals that end in:
    a. "ite" and "ide"
    b. "ide" and "ium"
    c. "ium" and "ate"
    d. "ide" and "ace"

39. This term describes how well the foam stands up to the heat of the fire:
    a. fuel tolerance
    b. knockdown capability
    c. burnback resistance
    d. heat resistance

40. This type foam is often used for subsurface injection:
    a. fluoroprotein
    b. protein
    c. synthetic detergent
    d. aqueous film-forming foam (AFFF)

41. This type foam is considered unsuitable for Class B fires:
    a. protein
    b. fluoroprotein
    c. synthetic detergent
    d. AFFF

42. This is the most versatile foam in use today:
    a. AFFF
    b. synthetic detergent
    c. alcohol-resistant aqueous film-forming foam (AR-AFFF)
    d. fluoroprotein

43. The most common method of proportioning foam is:
    a. batch mixing
    b. induction
    c. injection
    d. premixing

44. When using an inline eductor for foam operations, this is generally the maximum amount of hose that should be permitted after the eductor:
    a. 50 ft
    b. 100 ft
    c. 150 ft
    d. 200 ft

45. This method of foam application involves piling up finished foam in front of you and pushing or sweeping the foam onto the burning fuel:
    a. roll-on
    b. bank-down
    c. snowstorm
    d. push-back

46. To apply foam to a burning, running spill and attempt to contain it as well, firefighters operating the foam nozzle should:
    a. Work ahead of the spill.
    b. Plunge the stream into the spill to push it back.
    c. Work adjacent to the spill.
    d. Work behind the spill.

47. When the foam operation from an eductor begins to flow, what action should be taken first by the nozzle operator?
    a. Direct the stream in front of the fire to apply foam to the burning liquid.
    b. Direct the foam into the spill to push it away.
    c. Do not flow for 30 seconds to allow finished foam to build at the nozzle.
    d. Direct the nozzle away from the spill until foam flows instead of water.

48. Before any foam operation is begun, firefighters must ensure that:
    a. Multiple types of foam are on hand.
    b. A location for foam runoff must is designated.
    c. An adequate volume of foam concentrate must be on scene.
    d. A water supply officer is assigned.

49. Although excellent training tools, why are pit fires unrealistic when combating spill fires?
    a. Fuel spills are rarely as deep as a pit fire.
    b. The pits allow for a closer approach.
    c. There is no debris in the pit to complicate the operation.
    d. Personal protective equipment is not required for pit fire training.

50. The foam produced by this system has the ability to stick to vertical surfaces and has found acceptance in structural firefighting:
    a. around the pump
    b. inline eductor
    c. proportioning
    d. compressed air

CHAPTER 27 ANSWER KEY

Question #	Answer	Page #	Question #	Answer	Page #
1	b	905	31	d	926
2	d	906	32	b	927
3	a	907	33	c	929
4	c	907	34	b	929
5	d	908	35	a	929
6	c	909	36	b	932
7	b	909	37	c	933
8	a	911	38	b	933
9	a	912	39	d	934
10	c	913	40	a	934
11	c	914	41	c	934
12	c	915	42	c	934
13	b	915	43	b	935
14	d	915	44	d	936
15	b	917	45	a	938
16	d	917	46	a	938
17	b	917	47	d	938
18	c	917	48	c	939
19	b	918	49	c	940
20	d	919	50	d	940
21	b	920			
22	a	921			
23	c	921			
24	a	921			
25	c	921			
26	b	922			
27	c	922			
28	a	924			
29	c	924			
30	b	925			

Origin and Cause Investigation

by Richard S. Wolfson

1. Fire investigations often lead to changes in these guidelines:
    a. firefighter SOPs
    b. training mandates
    c. fire and building codes
    d. documentation guidelines

2. This is an integral part of the fire investigation:
    a. initial complement of apparatus
    b. firefighter's observations
    c. collapse damage to the structure
    d. prefire plans of the building

3. Which of the following will help the investigator determine the cause of the fire occurrence?
    a. Protect evidence by covering it.
    b. Ensure that the fire is completely out by conducting a thorough overhaul.
    c. Question the homeowner regarding the fire.
    d. Minimize overhaul operations.

4. When should the area of origin be treated as a crime scene?
    a. when the fire is determined to be incendiary
    b. at all times until it is determined otherwise
    c. when multiple fires are encountered
    d. when all information is not available

5. You have been sent to check on the utilities. When checking the electrical panel, you notice that several circuit breakers are tripped. What do you do?
    a. Reset the breakers and stand by in case they trip again.
    b. Do not reset the breakers, but make note of the information.
    c. Request that the incident commander take a look.
    d. Put tape over the breakers so they cannot be reset by unauthorized personnel.

6. Different types of heat energy that initiate a fire by causing a fuel to ignite are called:
   a. fuel sources
   b. sources of ignition
   c. energy sources
   d. pyrophoric initiators

7. A chemical or biological process that allows certain materials to generate heat and ignite on their own without the presence of an external heat source is called:
   a. self-sustained pyrolysis
   b. spontaneous ignition
   c. spontaneous combustion
   d. autoignition

8. The base of this type fire pattern points to something on the floor that has ignited at some point on the fire:
   a. light pattern
   b. inverted V pattern
   c. fingerprint pattern
   d. V pattern

9. If you were looking at an object that was crazed, you would be looking at:
   a. masonry
   b. steel
   c. glass
   d. wood

10. This type of electrical malfunction is caused when the insulation around two conductors in a wire is broken and the conductors come in contact with each other:
    a. arcing
    b. short circuit
    c. overload
    d. trip

11. Suppose you have just extinguished a car fire. You are retrieving tools, and as you reach into the passenger compartment, you notice a strong smell of gasoline. What do you do?
    a. Call for a hoseline to wash down the interior.
    b. Notify the police on scene.
    c. Check around the vehicle for a spill.
    d. Notify your officer of the smell.

12. An item that is used to spread fire from one area of the building to another is called a
    _____
    a. trailer
    b. flash-bang
    c. leader
    d. wick

13. A defective water heater or a fire pattern is an example of this type of evidence:
    a. testimonial evidence
    b. documentary evidence
    c. demonstrative evidence
    d. proprietary evidence

14. An insurance policy is an example of this type of evidence:
    a. testimonial evidence
    b. documentary evidence
    c. demonstrative evidence
    d. proprietary evidence

15. Witness statements are an example of this type of evidence:
    a. testimonial evidence
    b. documentary evidence
    c. demonstrative evidence
    d. proprietary evidence

16. When collecting evidence, and a sample of an item that will react with metal is being collected, instead of being placed in a metal can, the sample should be put into a:
    a. cardboard box
    b. manila envelope
    c. glass jar
    d. plastic jar

17. If a container must be transferred to a laboratory for testing, it must be signed off by the laboratory technician. This is called:
    a. legal transference
    b. chain of custody
    c. demonstrative transfer
    d. custodial procedure

18. Data in the form of facts collected by observation, experiments, or other direct data-gathering means are called:

    a. experience data

    b. empirical data

    c. researched data

    d. virtual data

19. When taking photographs of the scene, which area should be photographed first?

    a. the exterior of the structure

    b. fire protection systems

    c. all rooms and areas involved in fire

    d. the fire debris

20. This 1978 U.S. Supreme Court decision, *Michigan v. Tyler,* affirmed the fire department's ability to do what?

    a. Conduct investigations.

    b. Prosecute arsonists.

    c. Use photographs and sketches in court.

    d. Maintain control of the fire scene.

CHAPTER 28 ANSWER KEY

Question #	Answer	Page
1	c	833
2	b	834
3	d	835
4	a	835
5	b	838
6	b	836
7	b	836
8	d	837
9	c	837
10	b	838
11	d	838
12	a	838
13	c	839
14	b	839
15	a	839
16	c	839
17	b	839
18	b	839
19	a	840
20	d	841

Fire Prevention and Public Education

by Tom Kiurski, edited by Becki White

1. Among the many measures that can be taken to reduce fire losses, perhaps none is more important than:
    a. training firefighters in suppression techniques
    b. properly staffing apparatus
    c. educating people about fire
    d. documenting incidents properly

2. What do building codes regulate?
    a. where buildings may be constructed
    b. how a building is to be constructed or renovated
    c. what activities may be conducted in the building in terms of fire safety
    d. the fire-suppression requirements for the building

3. What do fire codes regulate?
    a. where buildings may be constructed
    b. how a building is to be constructed or renovated
    c. what activities may be conducted in the building in terms of fire safety
    d. the fire-suppression requirements for the building

4. At what point in the construction process is a building permit issued?
    a. when a business submits blueprints to a city for review
    b. once the building is completed
    c. when the city building code officials ascertain whether the specified building meets the applicable codes and ordinances
    d. when a final plan is submitted that is agreed upon by the jurisdiction

5. This is a legal inspection that should be completed annually or more frequently for high-risk hazardous occupancies:
    a. plans review inspection
    b. fire prevention inspection
    c. building code inspection
    d. retro code inspection

6. Suppose you were on a building inspection seeking permission to enter a building. Suppose that permission was denied by the business representative. What action could be taken in this case?
   a. Explain that the business does not have the right to deny entry for inspection.
   b. Request police assistance.
   c. Ask to have that denial in writing.
   d. Begin activities to secure a search warrant.

7. Once permission s granted, where does the building inspection begin?
   a. at the main entrance door, where the alarm panel can be inspected
   b. on the building's exterior, where a walk-around provides access and layout information
   c. on the roof, to get a panoramic view of the building and its surroundings
   d. in the below-grade area, so that utilities and suppression equipment can be inspected

8. A stairwell in a multistory building constitutes which part of the exit system?
   a. exit access
   b. the exit
   c. exit passage
   d. exit discharge

9. This is an independent organization that does testing, listing, and examination of items as they relate to fire protection and safety.
   a. Factory Mutual (FM)
   b. Underwriters Laboratories Inc. (UL)
   c. Insurance services Office (ISO)
   d. American Society for Testing and Materials (ASTM)

10. This is a voluntary inspection done by firefighters to help educate the occupants about the dangers of fire and how to be safer in their everyday lives.
    a. preplan survey
    b. fire prevention inspection
    c. building familiarization visit
    d. home fire safety survey

11. What is the major issue with basements when educating the public to the dangers of fire?
    a. housekeeping
    b. access
    c. inadequate ventilation
    d. lack of detection equipment

12. What is the safe distance that is recommended for combustibles to be kept away from any heating source (furnace, fireplace, space heater, etc.)?
    a. 18 in.
    b. 3 ft
    c. 5 ft
    d. 10 ft

13. What is the absolute minimum number of smoke detectors that should be in the home?
    a. one
    b. one per room
    c. one per floor
    d. two per floor

14. Why should smoke detectors not be placed in the corner of the room, but rather on a ceiling or high on a wall?
    a. Fires do not burn in room corners.
    b. Proper ionization will not occur in corners, affecting performance.
    c. Dust accumulation in corners will negate their effectiveness.
    d. Air flow is restricted in corners.

15. When should home smoke alarms be tested?
    a. daily
    b. weekly
    c. monthly
    d. yearly

16. When should single-station unit smoke alarm batteries be changed?
    a. monthly
    b. yearly
    c. twice a year
    d. every 2 years

17. Research shows that the_____ smoke alarm responds faster to flaming fires, while the_____ alarm responds faster to smoldering fires.
    a. photoelectric, carbon monoxide
    b. ionization, photoelectric
    c. photoelectric, ionization
    d. ionization, infrared

18. The majority of arson is what type?
    a. juvenile
    b. revenge
    c. arson for profit
    d. psychological imbalance

19. What is the most important part of the home fire escape plan?
    a. a map of the home showing escape routes
    b. the designation of a family meeting place outside the home
    c. location of the fire extinguisher and first aid kit
    d. the plan must be practiced at least twice a year

20. When doing a fire station tour, how should the tour begin?
    a. Ask questions to determine what the audience knows.
    b. Discuss the rules for the tour.
    c. Introduce the supervising officer of the station.
    d. Show the apparatus to the audience.

CHAPTER 29 ANSWER KEY

Question #	Answer	Page #
1	c	845
2	b	846–847
3	c	847
4	d	847
5	b	848
6	d	848
7	b	848
8	b	849
9	b	849
10	d	850
11	a	851
12	b	851
13	c	855
14	d	856
15	c	859
16	c	856
17	c	857
18	a	858
19	d	860
20	b	862

Vehicle Extrication

by Dave Dalrymple

30

Vehicle Extrication

1. Proper size-up of a motor vehicle accident scene requires a _____ approach.
   a. rapid
   b. methodical
   c. upwind
   d. highway

2. During size-up, vehicle orientation and observable damage may indicate:
   a. possible injuries and possible entrapment
   b. road conditions that led to the accident
   c. if additional roads need to be closed
   d. how many people may have been in the vehicles

3. After making sure the scene is safe to work in, the next step is to:
   a. Stabilize the vehicle.
   b. Make visual contact with the victims.
   c. Maintain verbal contact with the victims.
   d. Mitigate vehicle hazards.

4. Before removing the power to a vehicle, a firefighter should:
   a. Wash down the battery to remove any leaking acid.
   b. Determine whether all batteries need to be disconnected (if there is more than one).
   c. Wait for orders from the incident commander.
   d. See if power is needed for any accessories, such as power seats.

5. To completely remove power from a damaged vehicle, before the battery or batteries are disconnected, the firefighter should:
   a. Talk to the victim to find out the battery location.
   b. Determine the condition of the batteries, to see if they need to be disconnected.
   c. Shut off the engine and remove the keys from the vehicle.
   d. Flatten a tire to prevent the vehicle from rolling.

FIREFIGHTER II

6. At the accident scene, the line officer in charge of the rescue must:
    a. Ensure that the roadway is closed.
    b. Begin hands-on victim management.
    c. Devise a tactical plan of action.
    d. Set the strategic goals pertaining to life safety.

7. Most vehicles today are built around the concept of structural integrity, known as:
    a. full-frame construction
    b. unibody construction
    c. space-frame construction
    d. caged construction

8. The energy absorption area that has the greatest effect on how extrication will be performed is:
    a. roof of the vehicle
    b. passenger compartment sides of the vehicle
    c. rear of the vehicle
    d. front of the vehicle

9. In more modern vehicles, dashboard reinforcement has:
    a. been strengthened as a result of the use of high strength steels and alloys
    b. been weakened as a result of the use of plastics and lightweight material
    c. remained the same as in older cars
    d. been removed since the introduction of energy absorption areas

10. Dash displacement in a modern vehicle may include the added step of an additional cut in the fender in order to _____ the crush zone and allow the dash to _____ the rest of the vehicle.
    a. sever; move with
    b. access; be pushed forward from
    c. access; remain secured to
    d. sever; move independently of

11. Some modern day vehicles have more than _____ of high-strength steel and alloys within them.
    a. 25%
    b. 50%
    c. 63%
    d. 75%

12. Using a hydraulic rescue tool to cut vehicle sections containing polyester foam reinforcement may become difficult because:
    a. The material crumbles and falls away.
    b. Additional force is needed to compress and shear the material.
    c. Heat from the tool causes the material to melt and adhere to the cutting blades.
    d. Leaking hydraulic fluid falling on to the polyester foam will cause noxious fumes.

13. Another option for gaining access to the passenger compartment, if a door removal proves difficult, is:
    a. side removal evolution
    b. B-post tear evolution
    c. B-post rip evolution
    d. all of the above

14. All of the following describe motor vehicle glass except:
    a. laminar
    b. dual-paned
    c. polycarbonate
    d. enhanced-protective

15. When it is necessary to disconnect a battery in a damaged vehicle, the greatest concern for firefighters is:
    a. battery voltage
    b. the difficulty of finding the battery
    c. acid leakage
    d. knowing which battery cable to disconnect

16. Information on gasoline-electric hybrid vehicles can be found in the:
    a. *U.S. Highway Administration's Guide to Hybrid Vehicle Accidents*
    b. *U.S. Department of Agriculture Bio-Fuel Vehicle Handbook*
    c. *U.S. Department of Transportation Emergency Response Guidebook*
    d. *U.S. Department of Energy Vehicle Accident Guidebook*

17. When a hybrid vehicle is shut down and secured, there is _____ power in the high voltage battery.
    a. always
    b. never
    c. sometimes
    d. All of the above can be correct, because power amounts change from vehicle to vehicle.

18. After accessing the high-voltage battery in a damaged hybrid vehicle, the firefighter should:
    a. Cut completely through all cables to stop power to the vehicle.
    b. Disconnect all battery cables from the connection points.
    c. Disconnect only a single battery cable from a connection point.
    d. Leave the battery alone and not touch it.

19. In a hybrid vehicle, high-voltage battery components are _____ in color.
    a. red
    b. orange
    c. blue
    d. green

20. Disconnecting the battery will remove energy to all the supplemental restraint systems in a vehicle. This statement is:
    a. True; because all systems rely completely on battery-supplied power.
    b. True; because even mechanically activated systems need a small amount of battery power to be activated.
    c. False; because some side systems are only mechanically activated.
    d. False; because some front systems are only mechanically activated.

21. Supplemental restraint system (SRS) inflation devices may be found in:
    a. roof posts
    b. the roof structure
    c. the vehicle body
    d. all of the above

22. At the scene, the best way to determine if a vehicle is equipped with a side curtain air bag inflation module is to:
    a. Strip the interior trim before cutting.
    b. Cut through the exterior vehicle skin only.
    c. Cut the roof posts, and remove the interior trim.
    d. Refer to the manufacturer's Web site.

23. When conducting a size-up of a motor vehicle accident, if you discover the presence of new vehicle technology, you should:
    a. Tell the incident commander only.
    b. Identify the hazard and inform the rest of the your crew.
    c. Keep the information to yourself, unless the crew needs to work in the area in question.
    d. Note the information on the after action report.

24. During a front-side motor vehicle accident where the occupant is not using a seat belt, the frontal airbag tends to:
    a. Help the occupant maintain a seated position.
    b. Help eject the occupant out of the adjacent window.
    c. Push the occupant down and under the dashboard.
    d. Raise the occupant into the top of the windshield.

25. After rescuers evaluate a patient for airway, breathing, and circulation problems, the next step in patient care is:
    a. oxygenation of the patient
    b. C-spine management of the patient
    c. establishing communication with the patient
    d. to reevaluating the patient for injuries

26. Reproportion of the inside of a vehicle and strategic cutting operations help define the concept of:
    a. work-circle making
    b. compartment spreading
    c. dislodgement
    d. space making

27. At a motor vehicle accident, the most dynamic hazard facing rescuers is the:
    a. vehicle
    b. environment
    c. street condition
    d. potential for fire

28. Removed vehicle components should be placed outside the work area and with the interior facing upwards because of
    a. possible undeployed safety systems
    b. the limited number of sharp edges on the removed component being exposed
    c. the need for accident investigators to see the interior of the passenger compartment
    d. requirements to limit any additional damage to the components

29. The vehicle stabilization evolution using a strut with a ratchet strap to create one or more right angles against a vehicle is called:
    a. right-angled cribbing
    b. braced angle cribbing
    c. tension buttress cribbing
    d. step chock cribbing

30. When removing a roof from a vehicle, the roof post closest to the patient's head is cut:
    a. first
    b. last
    c. second to last
    d. It does not matter in which order the roof posts are cut.

31. Straight blade hydraulic cutters usually have two points of contact, and curved blade cutters will usually have _____ points of contact for cutting.
    a. three
    b. four
    c. five
    d. six

32. Because side impact and side curtain airbags can be affected by static electricity, as well as pressure and shock, any wires found in the extrication pathway during a door or roof removal should be cut using:
    a. hand cutters
    b. reciprocating saws
    c. hydraulic cutters with a curved blade
    d. hydraulic cutters with a straight blade

33. As vehicles construction advances with more hardened materials, which of the following tools will become more important for faster and safer extrications?
    a. reciprocating saw
    b. spreader
    c. ram
    d. cutter

34. With proper operation of the proper tools at an extrication scene, rescuers should strive for _____ of scene time.
    a. 20–25 minutes
    b. 15–30 minutes
    c. 10–15 minutes
    d. 5–10 minutes

35. Which of the following tools can be used to pull over a greater distance than a spreader?
    a. cutter
    b. combination tool
    c. lift bag
    d. ram

CHAPTER 30 ANSWER KEY

Question #	Answer	Page #	Question #	Answer	Page #
1	b	881	31	b	897
2	a	882	32	a	895
3	a	882	33	d	897
4	d	883	34	b	868
5	c	882	35	d	877
6	c	883			
7	b	885			
8	d	869			
9	a	869			
10	d	869			
11	c	869–870			
12	b	870			
13	d	870			
14	a	871			
15	b	882			
16	c	872			
17	a	882–883			
18	d	882–883			
19	b	883			
20	c	873			
21	d	881–882			
22	a	883–884			
23	b	883			
24	c	888			
25	b	889			
26	d	889			
27	a	883			
28	a	884			
29	c	886			
30	b	890			

Support of Technical Rescue Teams

by Larry Collins

1. A Firefighter II should have the ability to establish public barriers, retrieve various rescue tools, and:
   a. Assist as a member of the rescue team when assigned.
   b. Establish a rescue plan.
   c. Develop tool inventory lists.
   d. Recognize and request additional resources.

2. The acronym LCES stands for:
   a. loose cannon, easily scared
   b. lacks common emergency skills
   c. location commands extra safety
   d. lookouts, communication, escape routes, and safe zone

3. In a water rescue scenario, the term *river right* best describes:
   a. the right side of the river as a person looks upstream
   b. the right side of the river as a person looks downstream
   c. the side of the river opposite from where a person is standing
   d. any ground area adjacent to the right side of the river

4. The statement "deep water is slowed by friction with the bottom, while the fastest water is found near the surface riding over the top of the slower current below" describes:
   a. layered flow
   b. helical flow
   c. laminar flow
   d. diluted flow

5. Slower moving water along a shoreline will circulate with water in the middle of the river in a spiral motion known as:
   a. corkscrew flow
   b. helical flow
   c. laminar flow
   d. downward center flow

6. Areas of calmer water separated from the main downstream flowing current are known as:
   a. eddies
   b. hydraulics
   c. holes
   d. breakers

7. The statement "an artificial vertical or near-vertical drop structure that stretches across a waterway, resulting in a hydraulic from shore to shore" best describes:
   a. a boil line
   b. a strainer
   c. a low-head dam
   d. shore eddies

8. The single most versatile tool used for water rescue is the:
   a. personal flotation device
   b. the rescue board
   c. air-inflated hose
   d. throw bag

9. When a car that has crashed into a body of water begins to tilt vertically, the rescuers should:
   a. Attempt to tie the vehicle off to a solid object.
   b. Move far away from the car because it is about to submerge.
   c. Continue to work in and around the car to complete the rescue because the car is about to submerge.
   d. Understand that the vehicle interior will soon be at an equalized pressure.

10. For most firefighters involved in an ice rescue, the tactics include remaining on the shoreline and tossing a rope bag to the victim, attempting to the reach the victim with a pike pole, and:
    a. talking the victim through a self-rescue
    b. using a rescue board
    c. breaking the ice between the victim and the shore so the victim can swim to shore
    d. crossing the ice over a ground ladder to reach the victim

11. When operating at mud and debris flow incidents, the operational zone may be to within _____ feet from the edge of the flow.
    a. 50
    b. 75
    c. 100
    d. 125

12. The five stages to effectively manage the scene of a collapsed structure include size-up, surface search and rescue, void space search, general debris removal, and:
    a. triage
    b. selected debris removal
    c. logistics
    d. establishing communications

13. During the stage of surface search and rescue, which of the following tactics would be considered inappropriate?
    a. allowing heavy equipment vehicles to operate to assist in searching
    b. moving structural and nonstructural items by hand
    c. calling out for victims
    d. using basic shoring to prevent secondary collapse

14. While technical rescue trained teams operate during void space search operations, the main role of a firefighter is:
    a. operating specialized equipment such as search cameras and listening devices
    b. installing shoring
    c. breaking and breaching into void spaces
    d. support the operation by gathering and delivering equipment

15. Selective debris removal from the collapse area starts from the _____ and continues to the _____ in a process known as _____.
    a. top; bottom; grading
    b. top; bottom; delayering
    c. front; rear; removing
    d. left; right; removing

16. When conducting size-up at a high-angle rescue incident, the first due firefighters should try to determine which of the following?
    a. what the victim's predicament is
    b. if there is a vehicle extrication problem
    c. if the situation is best handled by a helicopter hoist operation
    d. all of the above

17. While awaiting the arrival of a technical rescue team at a potential jumper/suicide incident, firefighters should be able to do all of the following except:

    a. Prepare fall protection.

    b. Determine if specialized equipment is needed and request it.

    c. Provide scene security.

    d. Establish a medical group.

18. When called to assist at a marine emergency, firefighters should be prepared to do all of the following except:

    a. Work in and under the water to effect rescues.

    b. Conduct firefighting operations.

    c. Assist with hazardous materials operations.

    d. Provide medical care for injured victims.

19. On arrival at a trench rescue incident, firefighters should park their truck at least 50 ft away from the trench. The main reason for this is:

    a. to reduce the potential of vibration causing a secondary collapse

    b. to leave the area open for the next arriving apparatus

    c. to allow for an equipment staging area

    d. so there will be space to move other vehicles or machinery out of the way

20. After the original trench excavation collapse, there is greater than a _____ chance that a secondary collapse will occur.

    a. 40%

    b. 50%

    c. 60%

    d. 70%

21. A confined space that has a hazardous atmosphere, the potential for engulfment, and sloping floors best describes a(n):

    a. nonpermit confined space

    b. permit-confined space

    c. limited entry confined space

    d. extended entry confined space

22. First arriving firefighters at a confined space incident can quickly ascertain key information by:
    a. interviewing the site supervisor
    b. reading the entry permit
    c. obtaining any available material safety data sheets
    d. all of the above

23. All of the following are examples of indirect rescue tactics for firefighters operating at confined space incidents except:
    a. lowering ladders into the space
    b. lowering harnesses down to the victim
    c. entering the space to put SCBA onto a victim
    d. setting up a fresh air blower to ventilate the space

24. Which of the following initial actions for firefighters arriving at a "person-in-machine" type incident should not be undertaken?
    a. quickly running the machinery in a backward motion to free the victim
    b. conducting lockout/tagout procedures on all control devices powering the machine
    c. posting a guard at control devices, valves, or switches
    d. locating a person familiar with the system to provide technical advice

25. The first action to be taken by firefighters called to an elevator entrapment should always be to:
    a. Shut power down to the affected elevator car so it can be worked on.
    b. Turn the power on and off to the affected elevator car in an effort to get it working.
    c. Attempt to open the shaftway door with either a door key or hand tools.
    d. Have the dispatcher contact an elevator mechanic to respond.

26. The control for traction elevators will be found:
    a. adjacent to the hoistway, on the floor halfway up the building
    b. at the first-floor level of the building
    c. at the top of the elevator hoistway
    d. in the elevator pit below the hoistway

27. After opening an elevator hoistway door and finding that the car is just above the landing, the next step should be to:
    a. Open the elevator car door.
    b. Block the hoistway gap below the car with a ladder.
    c. Respond to the floor above, open the hoistway door, and climb down to the elevator, then attempt entry through the roof hatch.
    d. Begin communications with the occupants.

28. When working at an escalator emergency, after having determined that an escalator mechanic has been called, the first step before beginning operations is to:
    a. Engage the emergency stop button.
    b. Find and remove the landing plate.
    c. Locate the escalator control room.
    d. Start the lockout/tagout procedure where necessary.

29. Electrical energy to the escalator is provided by a motor usually located:
    a. beneath the lower landing
    b. in an escalator control room
    c. at the side of the escalator body
    d. beneath the upper landing

30. If a victim may not survive a long-term extrication because of an uncontrollable medical issue, or if a successful extrication may be impossible within the survivable limits of the victim's condition, an incident commander may need to consider calling the resources for a:
    a. helicopter evacuation
    b. police investigation
    c. surgical field amputation
    d. critical incident debriefing

31. When operating in caves and mines, rescuers may be subjected to:
    a. restricted entrances
    b. wall or ceiling collapse
    c. flooding
    d. all of the above

32. Safety regulations pertaining to abandoned tunnels come under the authority of the:
    a. Adopted standards of the National Fire Protection Association.
    b. Mine Safety and Health Administration.
    c. U.S. Department of Transportation.
    d. There are no accepted formal regulations.

33. If requested to respond to an incident of trapped workers in a commercial mine, firefighters may not be able to operate because:
    a. They lacked federal certification as a mine rescue team.
    b. They would be unfamiliar with the mine operation.
    c. They may have difficulty locating the site.
    d. The property owners may not allow them on to the site.

CHAPTER 31 ANSWER KEY

Question #	Answer	Page #	Question #	Answer	Page #
1	a	904	30	c	947
2	d	905	31	d	949–950
3	b	905	32	d	950
4	c	906	33	a	951
5	b	906			
6	a	906			
7	c	907			
8	d	909			
9	b	910			
10	a	911			
11	c	913			
12	b	924			
13	a	1932			
14	d	925			
15	b	924			
16	d	925			
17	c	927			
18	a	930–931			
19	a	930			
20	b	932			
21	b	934			
22	d	936			
23	c	937			
24	a	940			
25	d	943			
26	c	943			
27	b	946			
28	a	947			
29	d	947			